T0283556

HERE BEGINS
the DARK SEA

HERE BEGINS
the DARK SEA

Venice, a Medieval Monk, and the Creation
of the Most Accurate Map of the World

Meredith F. Small

PEGASUS BOOKS
NEW YORK LONDON

HERE BEGINS THE DARK SEA

Pegasus Books, Ltd.
148 West 37th Street, 13th Floor
New York, NY 10018

First Pegasus Books cloth edition June 2023

Interior design by Maria Fernandez

ISBN: 978-1-63936-419-0

10 9 8 7 6 5 4 3 2 1

Printed in the United States of America
Distributed by Simon & Schuster
www.pegasusbooks.com

To Sandra Quinn,

for a lifetime of support, laughs, and steady friendship

that deserves its own mappamundi.

Contents

We know only a small part of the innumerable things that Nature does, and those which seem to us to be unusual we do not believe. This occurs because Nature goes beyond the human intellect; and those who do not have an elevated intellect cannot grasp even the things of constant experience, let alone those that which are unusual. Thus, those who want to understand must first believe in order to then understand.
—Fra Mauro, mappamundi, 1459

Introduction

The first time I saw Fra Mauro's medieval map of the world, I had no idea it was significant, let alone a 550-year-old priceless masterpiece of cartography that was the Rosetta Stone of world maps.

Hanging there, all alone on a secluded space of wall outside the grand reading room of the Museo Correr in Venice, a spot where a tourist would usually just turn around and walk back to the entrance, it looked like an afterthought. I did know that the museum was connected to the city library, the Biblioteca Nazionale Marciana, which is full of old manuscripts and books about the city, but in truth, I had only walked into this hallway to find a passage down the stairs into the library. I might just as easily have walked quickly by, giving the map a cursory glance, but instead, I was arrested by its complexity and its beauty: I couldn't look away.

The map I saw was a towering circle of blue and white covered with busy writing, and it seemed alive. I was initially captivated by the blueness, obviously representing water, but this blue was also in motion with a flurry of whitecaps. A painter had seemingly put down a layer of navy blue pigment and then overlaid that saturated color with long strips of white wavy parallel lines, as if drawn with a fork. Unlike other maps that might show cresting waves on shorelines or maybe peaks of high water in a rolling sea, these waves were wiggling around every bit of open water while also containing the landmasses with their undulating dance. Interspersed among those waves were many tiny ships of various stripes. There were Chinese junks, Arabian boats with proud bows, double-masted schooners sporting billowing sails, large double-decker trade ships on their way to pick up cargo, and single-masted schooners with their sails held taut in what appeared to be perfect sailing weather. The land, painted in white, was also dancing. Edges of continents beckoned with fingers of lace, inviting the viewer to enter a cove, explore a bay, or go upriver to an inland pool of water. The land also sported readily identifiable topographical features: mountain ranges and deserts, as well as high cliffs and rolling hills, crawled across the surface; teensy tiny trees of various types littered the landscape; streams and rivers exuberantly wound their way up, down, and across the various continents, suggesting a multitude of possibilities for going inland. More striking were the human constructions that seemed to extend this map from the world of geography into the tale of human occupation. Humans make their mark, and their identity, by molding the natural world and they do so by building things. On this map, there were castles, towns, churches, arches, monuments, tombs, mosques, and towers, all exquisitely and accurately drawn for each culture, and correctly situated as representations of how humankind divides up the world in an attempt at ownership and identity.

The map also seemed to be calling out to me with a flurry of writing; across the map were handwritten notes which I could not understand—they were clearly not in Italian, and not even in Latin, as far as I could tell. In full paragraphs of tightly scripted text set down by a calligrapher's patient hand, all those words were trying to engage me in a conversation about this or that town, or river valley, or castle. Some lines were written directly on land, others on small white scrolls set into the blue water. The penmanship was in red, blue, black, green, turquoise, or brown, and the words and lines were perfectly spaced, as if typeset. It seemed obvious that those notes were explanations of what was depicted nearby, just as an encyclopedia uses both illustrations and accompanying text to describe an entry. I later learned they were written in an ancient version of Veneziano, the language of Venice, spoken and understood today only by Venetians and those who live in the Veneto region on the mainland. And yet, I could see that those inscriptions held the keys to this map. As an academic, I know references when I see them, but I wondered why a medieval cartographer felt it necessary to explain his work in such detail. I also wondered what those references were, where the cartographer got them, and who he was trying to impress or convince with all those words.

The writing, the rolling seas, the frilly land, and the busyness of human occupation on this map were held in place by a round three-tiered gold frame. That circular gold frame was, in turn, framed again by a large square of supporting wood with triangles in each corner, which also held small scenes and long inscriptions. Paradise with Adam and Eve was on the lower left, circular rings of what looked like the solar system on the top left, something that looked like the earth from afar at the bottom right, and a circle representing what I later learned were the four humors in the top right corner. How odd to have these existential issues booted off the map, outside the first frame. Standing there for quite a while, trying to get my worldly bearings by looking for the distinctive boot of Italy, I finally realized this map was drawn upside down. Much later, I would discover that the orientation of Fra Mauro's map was intentional, announcing that it was different from previous Western world maps. They were oriented toward the east, with Jerusalem in the center, to promote Christianity, but the creator of this map was a new breed of cartographer. He wanted to follow the dictates of science, not religion, and so he spun his world in a different direction.

There it was, inviting me to stand on my head to get a better view and talking to me in a language I didn't understand, and yet I wanted to get closer, to point a finger and trace those waves as they curved around the continents. Only later did I realize I had been staring at the oldest surviving medieval world map, a mappamundi, and that this creation is one of Venice's prized possessions, an artifact of such historical significance that it should have been under lock and key. But instead, this glorious piece of art, culture, cartography, and history was just hanging there for my personal viewing. I soon learned that this map, created by a Venetian secular monk named Fra Mauro in the mid-1400s, is both an accurate description of the known world at that time and a cipher to the past. It sits on the very transition of medieval knowledge born from wild speculation into Renaissance enlightenment that incorporated accurate observation and geographic reality. It was a groundbreaking cartographic and artistic masterpiece that urged viewers to see water and land as they truly existed. And yet, few have understood its significance.

For the next few years, while I was writing another book on Venice, Fra Mauro's map haunted me. For that book, *Inventing the World: Venice and the Transformation of Western Civilization*, I had researched and written about many maps and the contributions that Venetian sailors, traders, explorers, and artists had made. Venice is known for its maps, especially nautical maps called portolan charts that guided trading ships around the Mediterranean and out into the Atlantic. There is also any number of maps of the city of Venice, its lagoon, and the various islands that dot the water. These maps were used to show pride for a city known as La Serenissima (The Serene One) and to portray the might of the government of the Venetian Republic. But Fra Mauro's map was something quite different, and I wanted to know how it was made, how it fit into the life of the republic, what it meant about the transition from the Middle Ages to the Renaissance, and how it had affected the canon of human knowledge.

And so, I set out to learn everything I could about this map, the process of cartography, and world maps in particular. As a species that loves to do a walkabout, it makes sense that humans would have reason to invent mapping; surely the ability to record and communicate topography has always been an advantage for our kind. And yet maps have often gotten short shrift as modes of communication, taking a back seat to spoken and written language. And it's their visual impact that makes maps instantly and universally understandable, and relatable, no matter what language is used by the cartographer. Mapping is also an innate, instinctive skill. Every day we use our internal mental maps to decide how to go from one place to another, or to adapt to changes in our environment that interrupt the usual route. In other words, Fra Mauro's map pulled me into the idea of mapmaking and why humans feel compelled to make maps at all. I discovered that humans around the world, and throughout time, have always been mapmakers, and that we use maps as graphic representations of just about everything, and for all sorts of reasons. Drawing geography has long been a way to understand and guide, and it is an essential part of human nature. Ancient mappamundi are also typically filled with cultural knowledge, making them rich anthropological treasure troves. Investigating early world maps that came before Fra Mauro's also made me realize how important and compelling these drawings have been throughout human history. Any world map is the child of previous maps because geographic knowledge is cumulative, and often reactionary. On most of these older maps, there are religious icons and sites meant to reinforce or indoctrinate a particular spiritual identity and make for a religiously bound collative. Because I am focusing on Fra Mauro and his map, this means that I also needed to understand who he was as a cleric, a Venetian, and a cartographer of the Middle Ages. There is little written record of Fra Mauro the man, but the works of two scholars of this map, Angelo Cattaneo and Piero Falchetta, combined with my knowledge of the history of Venice and knowing the city well, allowed me to construct a picture of this quiet man who changed the world. I was also fascinated by the very process of making a map such as this, a huge map that covered the known world almost centuries ago before the internet, before satellite pictures, and even before the printing press and libraries full of books and information. I wondered where Fra Mauro got his information and who was on the team that worked on this map, and of course, why it was made in the first place. I also wanted to know about the materials used to construct this map and what it might have been like to write the inscriptions.

At its core, this map has two captivating levels: the geography, and the inscriptions. The geography is spectacular for its time and underscores various decisions that Fra Mauro made during its construction. For example, the fact that he eschewed much of Ptolemy, which broke away from the standard map of the Middle Ages, made him a cartographic revolutionary. And then there are the inscriptions. Since I had English translations by Piero Falchetta to rely on, it was fascinating to understand what was important, or not, to Fra Mauro as he made this map because he explained every decision in the texts. And it was during my reading of those texts that I came across the perfect title for this book—*Qui comenza el mar scuro*—which translated from Venziano is "Here begins the dark sea." It appears off the coast of the southern tip of Africa, floating in the Indian Ocean near Madagascar, an ocean that westerners still thought was an enclosed sea. What an apt warning for Western sailors entering that huge body of water after rounding the Cape of Good Hope, and for readers embarking on the story of an ancient world map. Investigating further, I discovered that various copies of Fra Mauro's map had been made, including one painted by a British artist and another a huge photographic reproduction that won a medal. Also, Fra Mauro's geography had a significant effect on future maps, which is why this map is held in high regard more than five hundred years later.

What I gained by entering the world of cartography through Mauro's mappamundi was a deep appreciation of how maps have defined our history and our species. They are part of our essential nature because we need them. First and foremost, maps are practical tools for getting from one place to another. But there are also many other kinds of maps with various functions that make them significant for our kind. They include topographic, political, cartometric, geological, road, nautical, cadaster, climate, and thematic maps, among many others. The various types speak to the usefulness of mapping, which, in turn, speaks to how humans love to superimpose facts and ideas onto a flat surface for visual representation. The verb *to map* says just that—to represent something visually by putting it spatially rather than in words, or to associate various elements with geography—or, even more simply, to place items in relation to each other denoting the connection or distance. Maps aim to make sense of the world and everything in it.

Maps also have other significant powers. They are symbols of identity, art, humor, ownership, and affiliation, as well as pictorial representations of history. Most notably, maps are political. They define countries and show unnatural boundaries that have been imposed on geography by people and governments, and that imposition changes how the world works. Maps are also often used as propaganda, to persuade or condemn.

Altogether, maps and mapping are integral to everyday lives. Today you might draw a quick map on a Post-it to explain to a friend how to get to a local coffee shop or look at a map that shows how many people are getting vaccinated in your city, or you might look up from your work and see a hand-drawn framed map of your town that you bought from a local an artist as a symbol of your attachment to where you live. Some might argue that we need paper maps less these days because of Google Maps and other apps that have moved cartography into the digital age for everyone. Those innovations are practical for driving, walking, or taking public transport, but they only cover one aspect of the many ways maps are used.

It is the combination of beauty and practicality that draws us to maps, and with new technologies in mapping, we have become even more personally involved in their creation and analysis. It takes no special skills to look at a map and pick out familiar places or wander into unknown countries and fantasize what it might be like to live there, to have one's place in nature shifted to the east, west, north, or south. Maps are guidebooks not just to other places, but to other ways of life, and Fra Mauro's mappamundi invites the viewer back into an older time when human society was just waking up and beginning to understand the world in all its magnificence.

CHAPTER 1

A Sense of Place:
The Human Urge to Draw Geography

The urge to map is a basic, enduring human instinct.
—Jerry Brotton, *A History of the World in 12 Maps*, 2012

*The map is thus both extremely ancient and extremely
widespread; maps have impinged upon the life, thought,
and imagination of most civilizations that are known
through either archaeological or written records.*
—J. B. Harley, *The History of Cartography*, vol. 1, 1987

A few blocks from me there is a map that I pass almost every day. It's about ten feet tall and twelve feet wide and pasted up inside the storefront window of a parking garage. It's not a display or an advertisement, nor does it have anything to do with parking. Instead, this map is a kind of street art put in place to keep passersby from looking into the offices of the parking garage. It shows downtown Philadelphia, with its clean grid of streets running north and south, east and west, and it's dotted all over with tiny churches, buses, buildings, violins, and binoculars, all of which denote various places to visit and the public transportation that can take you there. For example, the violin means music halls and the binoculars are on city hall because you can go to the top of that building and view the city. This map is also overlaid with lots of words—street names, place names, and neighborhood names. Overall, this window map is symbolic of the city; it represents, in miniature, the city and what it has to offer. Some maps—think subway maps—are graphically cleaned up for easy viewing, but this one is already very clean and straightforward because the streets of downtown Philadelphia do not twist and turn around features such as hills and streams because this is a planned city. Philadelphia was plotted out on empty land by William Penn and surveyor Thomas Holme in 1681 after Penn received the land from King Charles II. To Penn's credit, he then turned around and paid the Lenape for the land because they owned it. Penn and Holme designed their utopia as a grid of perfectly aligned streets with space for houses with gardens and many public parks because they intended this new city to be walkable, navigable, and simple. And they succeeded. Almost 350 years later, realized on the map pasted in a window, I can easily navigate as I walk or bike through the city according to their orderly design. I can also efficiently get around because the map of Philadelphia is now imprinted on my brain alongside maps of Davis, California, where I went to school; Ithaca, New York, where I lived for thirty years; Paris, where I spent a lot of time; and Venice, Italy, which I have learned to decipher over many visits. I can get around in every city I've ever lived in without a paper map or an app on my phone because there seems to be a personal atlas stored somewhere in my brain.

But there is nothing particularly special in my mapping abilities. Everyone does this all the time; we collect the symbolic representations of the landscapes we traverse, and they become etched in our brains after a few repetitions. We note roadways, rivers, landmarks, cities, mountains, lakes, and forests. We remember sidewalks, bus and metro stops, Starbucks locations, where our friends live, and how to get there. In some sense, humans can't seem to move without having a mental graphic representation of where they are and where they want to go. We could say humans are instinctive mapmakers because maps are essential for living a full life.

Drawings of place, that is maps, are items of communication, which means they reflect what one person wants to convey to another person.

But they are different from other forms of communication, such as language, which is based on sound and hearing, because maps are purely visual and pictorial. Maps are graphic outlines of whatever anyone can see, executed with some drawing implement, such as a finger covered in ocher pigment or a hand gripping a stick or stone to incise a rock, a blade of bone, a piece of wood, or a bed of sand. And the first maps didn't depend on language because written language, of course, had to be coded from the spoken word into letters and words first.

No one knows who drew the first map, or even what it contained, but throughout human history people have used the graphic of maps to show directions and to illustrate any number of subjects that can be placed, geographically, on the map. Maps, then, are also documents that provide information beyond directions. And that sort of exercise is called "mapping," meaning placing values across some terrain so that the viewer has a picture of the geographic distribution of whatever the subject might be—for example, income, housing type, jobs, crops, education, on and on. In effect, maps are the Swiss Army knife of communication. They tell a story, point the way, make an argument, dispel misinformation, and, most of all, give us a sense of place.

The ubiquitous nature of mapping suggests that diagraming our landscape is an ancient feature of human cognition and behavior and that we owe much of our evolutionary success to that ability. A map, in essence, is designed to convey information on a visual field and its purpose is to have that information easily absorbed without any verbal explanation. Mapmaking also takes a certain kind of cognitive ability that can translate three-dimensional topography onto a flat surface. That translation requires using miniaturized symbols that denote features. Mapmaking is the act of rendering what we see onto sheets of paper, computer screens, or other surfaces using coded and agreed-upon symbols that denote, in smallness, various features, and then placing those symbolic denotations in a context, or space. A whole map is one big symbol made up of many smaller symbols placed relative to each other in a way that makes spatial sense.

Without symbolic thinking, there is no map, and so when and why did humans gain this special ability of symbolic thinking that had to be in place before anyone could have drawn a map?

WHAT IS SYMBOLIC THINKING?

The simple definition of a symbol is an item that represents something else. The letters I write now, put together into words and sentences, represent my thoughts. They are connected to each other because one—the letters—represents the other—the words—but they are not the same because they are not exact duplicates of each other. That map of Philadelphia is not Philadelphia, but it represents Philadelphia, and I know it because my brain was designed to realize that fact. All maps are symbolic objects because they use graphics to represent something else.

Symbols are an alternative to a thing itself, and they are very handy, especially in communication. It starts with a signifier, the symbol that will be used to mean something else. The something else, the real thing that is being represented, is called "the significate." The signifier might look exactly like the real thing, or it might have nothing to do with what it is representing, but the collective, meaning lots of associated people, has agreed that the symbol will represent the real thing. Think of a stop sign, a cross, or an emoji. We have all agreed on what they mean and so we use those symbols to represent certain objects, feelings, behaviors, or events. Symbols are handy because they are shortcuts; they instantly and efficiently convey a deeper meaning. As such, symbols infiltrate all aspects of our lives. An analog clock, with its numbers and hands, shows time moving; a calendar displays the passing of days; a wedding ring announces someone is married; on and on.

Language is, of course, the granddaddy of human symbolism. With spoken and written words, words that are signifiers of something else, humans share everything. Letters and numbers are also perfect examples of the elegant leap from one bit of information into a wider communication universe. They are agreed-upon visual marks that, taken together in some order, tell a story. Letters and numbers quickly explain a situation without acting it out or demonstrating it in real time. Most importantly, symbols as representations influence the receiver and give deeper knowledge than the letter or number itself. They *mean* something.

What's even more remarkable is the fact that all humans use this symbolic shorthand.[1] Thinking symbolically is so much part of human nature that we usually don't pay attention to the process even when it's happening. For example, meeting someone new is a festival of symbolic thinking. We take in the other's clothing, accessories, and hairstyle and unconsciously work through what all that means about the person. Then we hear them speak and must symbolically interpret their words and the nuance of their voice that gives other meaning to what they say be it a joke or criticism. Those words are clues to the new person's mentality and listening to them talk is necessary to understand who that person is; we use the symbols they provide, including their words, to do it. Because of symbolic thinking, we can also refer to the past, explain the present, and imagine the future. We can create a parallel universe that represents whatever.

Along with language and numbers, maps are one of the most effective human uses of symbolic thinking. Throughout human history, maps have also been one of the main communicating devices of all peoples. They are also often richly symbolic beyond denoting landscape as they often carry the values of a culture, a nation, and the cartographer on their surface. Most often, a map is not just a map, meaning a map is usually so much more than directions. Historians have suggested that maps, as symbolic devices, came first and then may have proceeded very quickly to add written letters and numbers, suggesting they were the first carriers of written or drawn symbolic thinking.[2]

Maps also have a sort of immediacy with their graphic brand of symbolism—those tiny lines, squares, dots, and representative figures are usually easier to understand than reading a book full of letters or figuring out what columns of numbers might mean.[3] Show an illiterate person a map of their town and they instantly understand what it represents. Show an illiterate person a grocery list and they have no idea what those words, those symbols, mean. Historians have also suggested that maps were once the only good way to convey spatial information. Sure, you can tell someone to turn right and then left and look for the park on the right, but it's so much clearer, and more efficient, to draw a quick map.

Maps, then, are one of the earliest known modes of symbolic communication among humans. The very first maps probably came after speech, but before writing and numbers, and so maps represent one of the first uses of symbolism in communicating information.[4] Thinking symbolically is part of human nature, meaning it is a very old skill and one that must have been selected by evolution because it gave our ancestors an advantage.

THE EVOLUTION OF SYMBOLIC
THINKING IN HUMANS

Whenever a behavior or process is universal, it means that it has evolutionary roots and must have been selected for some reason over time. Across cultures, all humans are symbolic creatures and so that means the art of using symbolism must be somewhat encoded in our genes. Does that mean that we are born understanding symbols? Not quite.

According to the Swiss developmental psychologist Jean Piaget, the process of understanding and making sense of the world, including the value of symbols, is not really innate because babies are not born intuitively knowing that some objects can represent others. Instead, Piaget felt that symbolic recognition happens gradually over time as babies and children interact with their environment. He suggested that as little kids mature, they are always conducting experiments and gathering data by looking, touching, holding, falling over objects and people, and listening to what is being said, even if they can't respond with words yet. All that data gathering results in mental scenarios that make things fit together.[5] Small children also employ their active imaginations to sort out this cacophony of input into paths of learning, absorbing, and comprehending what the world is about. Through that intense process kids eventually realize that objects are real things and that they can personally cause things to happen, like throwing a ball, and that their self is separate from other selves. Eventually, language becomes a tool in development as others speak to babies and then as children talk themselves. Interwoven in cognitive development is the ability to know that things can represent other things, and that layer of cognition is essential to human action and thought.

Also, the fact that children acquire language simply by listening suggests that this fundamental skill, one that relies on the symbols we call words, demonstrates that humans are born with the capacity for developing symbolic thinking and that it sprouts early. Although Piaget felt that clear symbolic thinking doesn't appear until eighteen months to two years of age, recent evidence suggests that even newborns already have the concept that numbers represent something else, which is the very definition of symbolic thinking.[6] Piaget also claimed that there was a blossoming of symbolic thinking between ages two to seven, when children easily know that an object, such as a picture, or words in a book or spoken out loud, represents something else. This abstract ability can be seen when a child acknowledges that something exists even though it is not in the same room. In addition, symbolic thinking in children is on display as they make up stories and playact because that creative act means they know that what they are saying and doing represents other things.

The human mind, then, doesn't just acquire information like a blank slate waiting to be filled in. Instead, after birth, everyone enters an expansive mental world that is full of twists, turns, and hidden corners, where objects can have layers of meaning, and words, numbers, and pictures are the tools used to communicate about the past, the present, and the future.[7] In other words, we become ourselves through the medium of symbolic thinking as we interpret what others are communicating and use those very same agreed-upon symbols to respond.

This kind of symbolic thinking is part and parcel of being human, but it isn't especially unique to humans. Starting in the 1960s, various researchers decided to see if chimpanzees, our closest genetic relatives, were also capable of symbolic reasoning. Humans and chimpanzees split from a common ancestor about five to six million years ago and since chimps can't talk, it's difficult to know how exactly what and how they think. Some studies concentrated on language, knowing full well that apes do not have the anatomical ability to speak. So, researchers came up with protocols to teach the animals how to communicate with other types of symbols. There was success in teaching a chimpanzee named Washoe how to use 350 hand-motion signs of American Sign Language (ASL). Some of that gestural symbolic language she learned bit by bit through endless repetitive lessons with her trainers, but Washoe also picked up the meaning of untaught signs as researchers signed to each other. Eventually, she signed to other chimps and passed on her moments of symbolic thinking. This work showed that Washoe could navigate the world of symbols, because ASL is composed of hand and facial gestures that represent something else—that is—words.

Another chimp named Sarah learned how to equate plastic symbols with words. The researchers created plastic discs in various colors and shapes that had nothing to do with the object at hand. For example, they might use an orange square to mean a round red apple, and after many teaching moments, Sarah got it. She could also string symbols together and make new ideas.[8] A chimpanzee named Sheba was taught how to count, like we do, using Arabic numerals; after years of demonstration, Sheba could understand that the number three meant three things.[9] And schooled bonobos, the other kind of chimpanzee, can also use symbols to communicate with their human teachers.[10] Even more interesting, these symbolically educated chimpanzees sometimes used their learned symbols to communicate about food with each other in a way that was different than when the real objects were around, which signaled that they understood that the real item and the corresponding symbol were different but meant the same thing. The Japanese primatologist Tetsuro

Matsuzawa has also been able to teach chimps to understand the symbolic value of numbers and colors.[11] Other primates with smaller brains than chimpanzees are capable of some symbolic thinking as well. For example, South American capuchin monkeys can be taught to understand that a token represents something else.[12]

But an important point in comparisons with other primates is that they never use symbolic thinking in the wild, in their normal lives, as far as primatologists have observed. They must be schooled, over and over, and be positively rewarded with treats to understand that one thing can mean another thing. And so, these labor-intensive experiments show that while other primates might have the mental capacity to think symbolically it's not part of their normal repertoire. Since their behavior demonstrates a propensity for symbolic thinking if not the actual fact, our ubiquitious human ability to automatically use one item to represent another must have deep evolutionary roots. The hard-won evidence of this ability in other animals also shows how differently, and more advanced, this skill has evolved in our species. It takes a lot of time and effort to teach a chimpanzee or a monkey to pay attention and connect the dots but humans, even as little kids, pick up this skill easily, and then they use it repeatedly with the deepest understanding of what they are doing. All humans are born as cognitive sponges for symbolic meaning, and that's one reason they can draw and read maps, which are, by their very nature, symbolic objects.

SPATIAL COGNITION,
THE NECESSARY SPICE FOR MAPMAKING

Small children also quickly couple the ability of symbolic thinking with another animal-shared skill: an innate sense of space—the other essential ingredient for conceiving a map. Spatial cognition is the ability to take in information about what is around, make sense of it, and then navigate through that space. All animals have a sense of their position in space and where others are as well.[13] For example, many nonhuman animals know their territories and incise them with trails by following the same path day after day. Homing pigeons and domesticated horses know their way home; geese, salmon, and monarch butterflies make long treks north and south following routes embedded in their brains; many animals defend their home territories with calls, fights, and scent marking. Primatologists have also discovered that chimpanzees keep track of the various vegetation they eat in their ranges over time and have a clear schedule for revisiting them based on ripeness; chimps map their territory for resources, just as we do.[14] All these creatures know where they live and claim that land for themselves and they know how to navigate through it. Lacking this skill is inconceivable for any land, water, or aerial animal because to navigate the environment and find food while escaping predators is life itself. These animals, like human animals, also maintain ongoing mental maps of where they are, and they may be as good as humans at keeping that map in their minds as they traverse the area. In other words, we share with other animals that sense of space and place called spatial cognition that allows for mapmaking and map following.

Piaget and colleague Bärbel Inhelder were curious about how children—that is, all humans—acquire spatial knowledge.[15] A baby's world is awash with straight and curved lines, circles and squares, and empty spaces, and at first, infants are only aware that such things are there. Babies gaze around, taking it all in, and then they reach out and touch everything. From about four or five months of age, infants are holding objects and gathering spatial information as they grab and chew them. Through vision, touch, and mouthing, the infant brain begins to make sense of all those geometric relationships. Piaget and Inhelder suggested that the process begins with the baby cataloging the "proximity" of an object, that is, understating the difference between near and far. They start with objects near them but eventually take in the larger view and judge the proximity of things farther afield. Babies also soon know that some objects are separate from others and that there might be a spatial order among several objects. One might be behind another, some are enclosed by others, and little humans notice if things are continuous or not. By ten months of age infants pretty much have spatial relationships among objects down and know how to move things about and what happens when they do. In this sense, the baby is learning about the shape of objects as disconnected from self, and how those objects can be set within a larger space. So, babies very quickly understand proximity and separation, order, and enclosure, all features of spatial cognition.

Further studies by psychologists have shown that even when older children have never seen an overhead view of some landscape before, called the bird's-eye view, they absolutely understand that it is a map of some geography. They also readily recognize maps of their own village the first time they look at it in the form of a map.[16] Piaget felt the absorption of spatial cognition was not finished until middle childhood, when a youngster has lost their hold on their initial egocentric view of the world. After that switch, they also learn that things and people change size as they get farther away. At that point, kids have gained an understanding that how we size things in the real world is shaped by proximate distance from our eyes. They "get" perception, which is a spatial skill. Once in middle childhood, older children seem to have enough spatial reasoning to be able to give directions to another person, orienting those directions to the viewer, not him or herself. At that point, children are fully rooted in accurate spatial cognition.

This stage of understanding also means young humans start to internalize cognitive maps, which means they can picture space in their minds and whatever is in that environment, even if they look away.[17] The internal cognitive map directs our movements and allows us to "picture" a landscape or a town even if it's not drawn or etched on a surface sitting before our eyes. In that sense, physical maps are representations of cognitive maps. Other animals don't need physical copies to guide their way and for much of their daily travel and we also don't need physical maps in familiar places. The imagined cognitive maps we use all the time are based on the expectation that things operate in a certain way and will continue to do so. The route of one's morning routine—bathroom, kitchen, out the door to the bus stop—for example, is an imprinted cognitive map that has been internalized through repetition until it becomes an expectation or habit. It seems that the mind takes in information about the body's surroundings and what is happening, then codes it and stores it ready for recall. In fact, we are constantly mapping our personal environment, second by second, as the eyes snap a mental picture and load it onto the hard drive of the brain. The input is not just visual; cognitive maps are also founded on smells, sounds, and the physical relationship between the individual and everything else.

Cognitive maps keep track of our world, and in that tracking, they provide a sense of "normal." Cognitive maps also give off a feeling of mental safety; when the mind is skating through a known cognitive map there is less anxiety about the present and the future. Mental maps are also the "routes" that the mind takes, just like following a road map to a destination. Such routes, or frames of reference, can overlay social interactions, workflow, and home routines. These maps are not pictures per se, but fluid thought processes that follow a regular pattern and thus become ingrained. Not following that cognitive map, or circumstances upending some normal routine and thus an expectation, can be upsetting and anxiety-provoking. We run our day through unconscious cognitive maps, which have been established through the filters of symbolic thinking and spatial cognition, and those maps make up what each of us thinks is "normal life."

People aren't born knowing all this, but we are wired to absorb spatial information and the values that symbolic thinking brings and keep it forever. Altogether, these mental acquisitions create the values of space for a child as they engage in the real world and try to make sense of it, and then this information develops into cognitive maps which become the roadmaps of our lives.

WHEN DID HUMANS START TO DRAW MAPS?

Humans have taken the skills of symbolism, spatial cognition, and cognitive maps and rolled them into physical maps, which is one of our greatest, and most efficient, modes of communication. No one is sure exactly when humans gained this ability, but anthropologists have long been interested in when exactly the human lineage began to show the kind of symbolic thinking that could lead to the impulse for mapmaking.

Some have suggested that when our ancestors began chipping rocks into stone tools, they must have had a symbolic notion of what they wanted to produce and why. But there is no indication that making a tool is an act of symbolism. Sure, it means being able to envision an object and make it, but that act is more practical than symbolic. Also, the long record of human stone tools shows that there is no requisite connection between the manufacture of tools and any kind of mentality or brain size. Ancestral humans have had tools for about three million years, but there is no reliable evidence that they had symbolic thinking for that long. And so, many anthropologists believe that symbolic thinking appeared after anatomically modern humans left Africa hundreds of thousands of years ago, but that hypothesis might also not be right.

Our kind has a long history of evolving in Africa and then moving out of that continent in waves, presumably in response to environmental crises that made it necessary to look further afield for food. Those waves correspond with anatomical and cultural changes that also chart the evolution of our species from a near ape to the people we are today, and they make for an organizing system for understating our past. Exactly when the transition to fully modern humans, with all their cognitive skills, began is disputed. The traditional date for this final wave that started in Africa has been usually about two hundred thousand years ago; we have well-dated skeletons and lots of cultural artifacts to know these creatures looked just like modern people and had begun to have sophisticated cultural lives. Humans have not changed anatomically since that time, which is why we call these creatures "fully modern humans" or "anatomically modern humans" and label them *Homo sapiens sapiens*. These modern ancestors also experienced a flowering of symbolic thinking, expressed by the way they lived, what they constructed, their tool kits, how they used cultural artifacts, and their artistic expressions. We know they had symbolic thinking because the evidence is all over the place in their home bases, tools, art, and evidence of social behavior.

But recent discoveries in Kenya suggest that the transition to sophisticated symbolic thinking might have happened long before that transition to anatomically modern humans, or that the date for anatomically modern humans should be pushed back. The inhabitants of the upper layers of a site called Olorgesailie, in the African Rift Valley, dated over three hundred thousand years ago, apparently made finely crafted tools far superior to previous human ancestors, *Homo erectus*, who were the masters of the Acheulean hand ax, which they clutched for a million years. The Olorgesailie residents also used a variety of pigments and possibly used those paints for decorative purposes. Their stone tools were made from rocks sourced at other places, which might suggest trading, a sophisticated social skill.[18] Although there are, as yet, no fossil bones associated with these finds, the accoutrements suggest that whoever these early humans were, they had already evolved into savvy symbolists.[19]

More clear evidence of early symbolic thinking has been discovered at Blombos Cave, east of Cape Town, South Africa, dating from one hundred thousand to seventy thousand years ago. There, paleontologists uncovered carved beads made from marine tick shells, some of them carefully perforated, colored, and polished. And there is evidence that the beads had been strung together in a pattern. Such crafting suggests that Middle Stone Age peoples were using them for adornment and decoration, which in turn might mean a need for identity, a clear symbolic effort.[20] More beads have also been unearthed at a site far from South Africa, in Morocco, dated around eighty thousand years ago.[21] Etched ostrich egg bits were also discovered at a rock shelter in South Africa called Diepkloof, and they date from one hundred thousand to fifty thousand years ago.[22] Paleontologists at Blombos also found a piece of ocher, a naturally occurring earth pigment that comes in various colors, with cross-hatched marks intentionally cut into it. More telling, they found a small piece of stone with lines intentionally drawn in strokes with red ocher pigment. The researchers suggest, based on their analysis of the lines, that the inhabitants used a pointed ocher "crayon"; that is, an artistic tool purposefully fashioned to draw those lines. That piece of rock from seventy-three thousand years ago is considered the very first human drawing.[23] They also found a paintbrush made from animal bone and shells that had been used as paint pots to make the tints of red ocher and charcoal.[24] The researchers also call the lines "graphic designs," which means they might represent the beginnings of the ability for spatial representation, and even the first sign of human mapmaking. Taken together, these pieces of evidence suggest that anatomically modern humans were thinking symbolically at least one hundred thousand years ago, if not before.

For all we know, symbolic thinking might not even be the sole provenience of anatomically human ancestors. That idea has come to a head in recent years because of a discovery in southern Europe and the possibility that our human cousins, the Neanderthals, might also have had symbolic thinking. Between sixty thousand and thirty thousand years ago, two types of humans inhabited that region, although we have no idea if they interacted with each other on a day-to-day basis. One type is colloquially called Neanderthals, and they were there first. Later, in yet another wave from Africa, anatomically modern peoples arrived in Europe. These modern humans are sometimes referred to as Cro-Magnons. Although the two types of humans lived at the same time and in the same place, they were anatomically different, enough that paleontologists have given them different subspecies names. Neanderthals are officially *Homo sapiens neanderthalensis*, while our direct lineage bears the name of humans today, *Homo sapiens sapiens*. It seems that Neanderthals were a geographically restricted group of individuals who did not spread throughout the world like so many of our other forbears. And they disappeared about thirty thousand years ago while their neighbors in Europe, the fully modern humans, continued. The ongoing controversy about these two types revolves around the issue of symbolic thinking, which is expressed dramatically in the drawings and paintings on cave walls. Neanderthals were culturally sophisticated creatures—they buried their dead—but paleoanthropologist Ian Tattersall still feels they had a more rudimentary form of what we now consider symbolic behavior. "There is little if anything in the Neanderthal record that can be convincingly seen as intended to represent anything in the external or abstract worlds," he writes.[25] But other paleontologists have found what they consider clear symbolic thinking by Neanderthals at various sites. For example, at ten Neanderthal sites across Europe and Siberia archaeologists have uncovered raptor talons with cut marks made by the hands of Neanderthals and they conclude that the bird claws could only have been used as body ornaments, like beads.[26] Beads that have been cut from shells and decorated with ocher pigment; cave wall drawings, including hand stencils; and evidence of body paint have also been found at other Neanderthal sites.[27] The dates of some of these sites show that Neanderthals were there long before the appearance of fully modern humans in the area.[28]

Neanderthals had the same brain size as modern humans, and if brain size is any measure of symbolic thinking, they could have had that ability. Perhaps the idea that only fully modern humans were sophisticated thinkers follows old, self-aggrandizing, notions that modern humans are something special, smarter, and more creative than their forbears.[29] But whatever cognitive power fully modern humans had over Neanderthals, they seem to have overtaken these evolutionary cousins and been a factor in wiping them out; maybe our direct ancestors conquered all the Neanderthal territory and chased their competitors into extinction. Underlying this argument about Neanderthals versus anatomically modern humans is the suggestion that the sophisticated symbolic behavior of modern humans made for a superior intellect, which in turn made for better strategizing, communicating, and organizing, and perhaps that combination was the key to humanity's success over others who looked a lot like them but didn't have the advanced superpower of elaborate symbolism, just the rudiments of that possibility.

No matter the roots of symbolic behavior, it is clearly in evidence around forty thousand years ago in the many cave paintings in southern Europe and across the globe, such as on the island of Borneo (Kalimantan, Indonesia).[30] The consensus is that this elaborate and complex art and other crafted objects such as jewelry and figurines were conceived and executed solely by fully modern humans, not Neanderthals, nor any other ancient lineage.[31] Drawings such as these continued through centuries. Ancestral humans seem to have been compelled to write on walls, adorn themselves with trinkets, and fashion figurines along with their weapons. In that making and artistry, we see a heavy layer of symbolic thinking as well as the beginnings of the kind of complex culture we have today.

Cave art in particular tells the story of our symbolic history. Standing in one of those caves and staring at the human figures, animals, and graphic squiggles made by early humans, as I have done at Lascaux cave in France, is a spiritual experience. The symbolic nature of the work is objectively clear. But it's the human connection between whoever made those marks so long ago and the living person standing there today looking at them that is so transcendent; in an emotional, psychological, and intellectual sense, we understand these pictures today and feel them in our souls because we share the same symbolic brain with our Cro-Magnon ancestors. Beyond the artistic value, cave art also includes many geometric shapes and hatched lines that seem to present some sort of story, a cognitive map of an event such as a hunt, and it's always possible that these scenarios were initial attempts at mapmaking.[32]

THE VERY FIRST MAPS

The oldest recognizable map in the world comes from Ukraine and is dated as much later than cave art, around fifteen thousand years ago. An etched mammoth tusk was unearthed along with many mammoth bones at a site called Mezhirich. The scratched lines on the tusk are not random but form a drawing or picture that is likely some kind of map. The design is made with lines, some straight, some zigzagging, presumably to denote a topographical feature. There seems to be a river at the bottom of this first map, and marks of a mountain or hill have been placed at the top. In between are more small lines that look like houses or huts. All the slashes are lines; none of them are curved in any sort of artistic way. There are also no animals and no human figures. This tusk is markedly different from any cave painting in its lack of story and figurative representation. Then, there is its straightforward presentation, which also makes it seem more map-like.

There is a clearer ancient map at the archaeologically famous site Çatalhöyük in Turkey. Çatalhöyük was a thriving city with complex cultural traditions for over two thousand years, as the several layers of buildings denote.[33] Besides the many murals painted on the walls of that city, there is also a map dated around 6,200 BCE. This map is painted on a wall and is almost 10 feet (3 M) long. It portrays rows of houses represented as black boxes with white squares and dots that might be windows or other decorations. Their placement right next to each other resembles the layout of actual houses that archaeologists have unearthed at Çatalhöyük with the entrances typically not at ground level but through the roof. Overlooking all the houses is a mountain painted orange with black spots and lines exploding off the top. Presumably, this part of the drawing represents the volcanic mountain Hasan Dagi, which still stands over this area today, although it has not erupted in 7,500 years or so.[34] This painting was a mural in someone's house, just as we now hang maps as decorative art in our modern dwellings. It's also the first map made from a bird's-eye view, which suggests a kind of sophisticated and abstract symbolic thinking that typifies modern humans. The artist was painting close-up but also imagining his or her city from afar and from above, a special cognitive skill that is necessary for mapmaking.

Perhaps the best-known ancient map is carved on a flat rock in the Italian Alps near the tiny city of Pescarzo. This inscribed drawing is part of a field of petroglyphs, now part of an archaeological park, where early humans spent a lot of time scratching out scenes on rocks. This one, called the Beldonia map, is of interest because it seems to be covered in images of what surely are houses, roads, fields presumably for growing, and topographical features such as streams. Superimposed on what is surely a map are some armed figures, a few animals, and one little house. This rock is 30 feet long (9 M) and 13 feet wide (4 M) and dates from about 1,000 BCE.[35] Scholar Christina Turconi also believes that the Beldonia map might be more than just a simple map. She acknowledges that it was carved at a time when personal property was becoming the norm and a map of a city might therefore be important, but Turconi also feels that this rock art had mystical or religious meaning and that the scene was also carved to protect the people who drew it.[36]

Unfortunately, no one knows the mindset of the artists and cartographers making these early maps.[37] Were these individuals simply recording some moment in time, and, if so, can we speculate why? Were they intended as mnemonic devices, to aid the creator in finding their way back or to a particular place once more? Or were they trying to communicate with others using graphics, perhaps telling a story, or giving instructions? It's also possible that some early humans were regularly drawing maps on surfaces that haven't survived, such as sand and dirt. These people were used to migration, following animals to kill and eat them, and mapping out a territory in their heads as they tracked the ripeness of fruits and other vegetation, and they may have made maps on temporary surfaces that have left no trace.

Those ancient humans who might have scratched directions in the sand or carved lines on wood were the first to practice the art of symbolic representation in the form of a map. From that point on, our ancestors all over the world scratched marks on walls and rocks, dipped their fingers into pigments, and gorged bones, shells, and horns to make marks of where they lived or where they were going. Or perhaps they were making those maps to explain visually and graphically some dream landscape or a long-ago event while placing that story into its topographical context.

MAPS ACROSS CULTURES

If drawing and reading maps is a universal human endeavor, then there should be contemporary cross-cultural evidence that people from various traditions and lifeways use maps.[38] That suggestion should work not only for economically advanced countries and cultures but also for Indigenous populations, especially those with traditions that involve the need for navigation across featureless places such as deserts or open water. Forest dwellers can plot out areas with paths anchored by obvious markers such as broken branches, piles of rocks, and distinct changes in topography. In contrast, Arctic and other ocean-going peoples such as Pacific Islanders have had to excel in their navigational skills and exercise their symbolic thinking and spatial cognition to figure out and record how to get about in those environments.[39]

So far, no one has done a broad anthropological comparison of contemporary Indigenous maps as clues to the development of cartography, so we have no good record or analysis of what might be called aboriginal maps and their place in the progress of human mapmaking.[40] But Indigenous people from all over the world with all kinds of lifeways do make drawn or carved maps.[41] Of course, every group needs to know where they live, what land is theirs and what belongs to others, and how to get about. Living in hunting-and-gathering groups, or as nomads, also means passing that information along either orally or visually to others and descendants. But for many cultures, there is no mapmaking tradition because they do not need maps. Maps are, after all, devices of communication, and if everyone already knows what the village looks like and how to get around, and there is no need to venture beyond home, there is no real need for a painted, drawn, or carved map. As historian of cartography David Woodward puts it, "Not until the demands of agriculture, private property, long-distance trade, militarism, tribute relations, and other attributes of redistributive economies transformed the discourse environment in which these firecrackers exploded was the light they emitted apparent. But then maps must have seemed the answers to prayers (why hadn't anyone thought of them before?)."[42]

The most well-known Indigenous maps are the stick charts from the Marshall Islands, which were first noted by Westerners in the nineteenth century.[43] The Marshall Islands consist of twenty-nine atolls and five islands arranged in two parallel lines. The Marshallese have always been great navigators around their various atolls, but they also ventured far away across the open ocean to other island groups. Their navigational points of reference were the stars and the wave patterns made as ocean water wiggled among their atolls. These Marshallese also made models of routes among their islands using sticks held together with bits of plant twine.[44] Each one is different in size and configuration, and therefore scholars believe that the variation reflects different makers.[45] But overall, they have a basic representation and so can be considered maps.[46] Some explain wave patterns in relationship to islands and atolls while others show general sailing or rowing directions. The large stick forms weren't taken on voyages but instead used to teach others before they set out. Smaller versions often went along in canoes, like a paper map in the glove box.[47] Today, these stick charts can be found in museums as ethnographic material goods no longer in use today but representative of a culture before Western contact.

The Kalaallit Nunaat Inuit of Greenland use driftwood to carve notches that represent the coastlines, islands, and the waters between them. These wooden maps are hand-sized and can fit into a mitten on a cold Greenland journey.[48] Right before colonization by conquering powers, Canadian and Alaskan Inuit also interacted with various expeditions by Westerners and contributed to the colonial maps, and they were highly accurate in their drawings.[49] Sometimes the Inuit drew maps on snow and sand, or even in the air, as they tried to help Western explorers find their way. And they did so over a hundred years of intrusion into their land.[50] Canadian Inuit also sometimes carved land features and their coastlines on walrus tusks.[51]

There is one early report of Australian Aborigines of the Torres Strait using rocks to lay out a plan of their village and the surrounding area and using that map as a teaching aid for youngsters learning group traditions.[52] In general, Australian Aborigines find their way by the stars and landscape features. These routes, some that crisscross Australia, are also remembered as "songlines" or "dreaming tracks."[53] These songlines are oral maps that also contain important information and they often appear in Aboriginal art as paths of dots and lines; they also figure in other levels of Aboriginal cultural traditions. These traditions are so closely held that only an Aborigine has access to their use or meaning. Although songlines are not maps in the usual way Westerners envision or comprehend maps, they are full of meaning and symbolism as well as directions communicated among people. The integration of directions for getting from one place to another into the spirituality, cosmology, traditions, and artistic creativity of Aboriginal peoples underscores how integral mapmaking is for an understanding of a people's place in a larger and more spiritual world. How Aboriginal peoples realize songlines into their art also foreshadows and echoes how medieval cartographers created maps of their world full of scary monsters, fantasized land and water, and places from imagined worlds such as the Garden of Eden. Both for Aboriginal people and medieval cartographers, maps were not so much about getting the landscape down on a permanent surface as they were about exploring the topography of the mind.

There are probably more examples of mapmaking in non-Western cultures, but they are presumably maps drawn on dirt, sand, and paper that were tossed away. Knowing the landscape is essential for people who make their living by herding, hunting, and gathering, especially over bare landscapes, and they are just as essential for groups who defend their territories. But the ubiquitous nature of maps means they are not always precious items that are preserved for posterity. Maps are usually practical objects, and often ephemeral, and therefore lost or tossed.

All humans are pretty good at wayfaring—that is, following a path and finding their way somewhere. Unconsciously, every time we step outside or get in the car, our minds are opening the mapping app in our brains and subconsciously figuring out which way to go. And so, it makes sense that there would be a long record of maps that picture an area where a group lives or depict a well-used trail from one place to another. Beyond the practicality of maps as devices for communicating routes or establishing ownership of land, there is another side to mapping that also interacts with how the human mind works, and that's the move to make maps that contain places far away and as-yet unknown. Humans are drawn to that too. They make maps of the sky, maps of the universe, and maps of the world. These cartographers are not so interested in communicating how to get about and are more interested in understanding the broader cosmology we inhabit. Such maps are existential, and they speak to our curiosity about how humans fit into the history of life and our universal need to believe we are living in an ordered and known place. In turn, these maps also say so much about what people are thinking during the time when that map was drawn; they speak of culture, identity, exploration, economy, and knowledge.

Once humans understood that the world was larger than one village, one country, or just one continent, the examples of maps of the world became historical flashcards that explain, even today, how people thought about life on Earth during the long history of our species.

CHAPTER 2

Mapping the World
Before Fra Mauro

*A world view gives rise to a world map, but the world
map, in turn, defines its culture's view of the world.
It is an exceptional act of symbolic alchemy.*
　　　　　—J. Brotton, *A History of the World in 12 Maps*, 2012

*The Mappamundi was an encyclopedia, and, like the
encyclopedias on CD-ROM, the form necessitated
abridgment. The advantage of the form was that
one could see the whole world at once, the medieval
equivalent of earth from space, or a "god's eye view."*
　　　　　—Evelyn Edson, *The World Map, 1300–1492*, 2007

*For my part, I cannot but laugh when I see numbers of persons
drawing maps of the world without having any reason to guide
them; making, as they do, the ocean stream to run all round the
earth, and the earth itself to be an exact circle, as if described by
a pair of compasses, with Europe and Asia just of the same size.*
　　　　　—Herodotus, *Histories*, 430 BCE

*There can be no comparison between one who lingers among
his kinsmen and is satisfied with whatever information
reaches him about his part of the world, and another who
spends a lifetime in traveling the world carried to and fro
by his journeys, extracting every fine nugget from its mine
and every valuable object from its place of seclusion.*
　　　　　—Abū al-Ḥasan ʿAlī ibn al-Ḥusayn al-Masʿūdī,
　　　　　10th century

On December 21, 1968, the crew of Apollo 8 lifted off from the newly minted Kennedy Space Center on Merritt Island, Florida, and headed for the moon. In the previous seven years, the United States and the Soviet Union had collectively conducted more than two dozen human space flights, but none of them ventured farther than 739 nautical miles from the surface of the earth. The goal for Apollo 8 was to go much further, nearly 250,000 miles to Earth's nearest celestial neighbor, the moon. They would orbit the moon but not touch down.

Crammed into a small space capsule and shot off their home planet into the universe, astronauts Frank Borman, James Lovell, and William Anders became the first humans to travel beyond low-Earth orbit, the first to orbit a celestial body other than Earth, and because of their historic vantage point, the first to see an earthrise. That moment was captured by Anders on December 24, Christmas Eve back on Earth. His iconic color photograph shows a slice of the moon in the foreground and the perfect sphere of the blue-and-white Earth as an independent body floating in space, lit up by sunshine as it emerges from the shadow of the moon. Sure, other space travelers had seen portions of the earth before as they orbited our planet a few hundred miles overhead, and there were full-Earth images taken by weather satellites, but nothing, absolutely nothing, could match Anders's image—the perfect sphere, the brilliant blue of the oceans, the wispy cloud cover, and all of it suspended in space.

It was the first time our species had traveled far enough to glance back and get a full view of home, the place where six or so million years ago human-like creatures stood up on two legs and began to explore their surroundings. During the following millennia, our kind then doggedly trudged to every corner of the planet looking for food and shelter, marking territories, developing cultures, evolving big and complex brains, mating and passing on genes, and becoming creatures who have a driving urge to understand where we live.

Anders's photograph was the culmination of that six-million-year quest to understand the world, our planet, as a geographic phenomenon. And yet people have, for a very long time, been trying to imagine and draft what they thought the world looked like without seeing it from space. Ever since humans discovered how to take up an implement and scratch on wet clay, people have tried to sketch the shape of the earth, how big it is, where landmasses are located, how seas and rivers and oceans surround and run through the land, and who lives where.

One of the great gifts of human consciousness is the ability to imagine and represent what one has not directly seen, and humans put that ability to work while mapping their world, even when we had no idea what the world really looked like until we first saw it floating in outer space.

HUMANS MAP THEIR WORLD

Among the many maps created by people over time and across cultures, one mode stands out as the most imaginative and creative, and the least practical—the world map. These maps don't show the way to get home, guide a traveler, or even inform accurately what belongs to whom. World maps are purely artistic in that they have always been made for grand effect. And it's their existential nature that makes these maps so captivating and intriguing.

Mappamundi are also products of their individual times.[1] These all-encompassing world maps chart the history of geographic knowledge during the year they were made, and they are evidence of what the various mapmakers knew and felt was important. These sweeping, impractical showpieces also echo the society in which they were produced; they are talismans of culture.[2] Maps of the world, then, are touchstones of humanity that make them more important, and more telling, than the more practical and less grandiose maps that simply diagram roads and cities.[3] Once people knew how to draw and write, and once there were permeant surfaces such as vellum and parchment to write upon, world maps became the storytellers of human experience.[4] Their story is our story, and that's why they matter.

There are all sorts of ways that the history of world maps could be organized.[5] They might be cataloged chronologically, by continent or

society, by style, if they were part of a manuscript or not, or even by
common features they might contain. In order to orient ourselves and
better understand the story of Fra Mauro's mappamundi and its creation completed in 1459, I am organizing the history of world maps
both chronologically and by source or origin—that is, the country or
affiliation of the mapmaker. Within this chronology, I organize these
maps by culture or society, which demonstrates how expansive the
push for images of the world has been, by all peoples ever since humans
could draw.

Fra Mauro did not make his map in a vacuum. Instead, he stood
on the shoulders of a long line of mappamundi cartographers who
had, for centuries before Fra Mauro first took up a paintbrush, made
decisions about how to represent the world. They, too, relied on what
others before them knew, and so the history of world maps is really one
map built upon another. Of course, it's impossible to know what Fra
Mauro knew about the various world maps I present in this chapter, but
surely he was familiar with other maps made in Italy, and perhaps also
with those made in other countries close by, such as Spain, Portugal,
and Germany. Given the nautical traffic to and from Venice because it
was a trading nation, he might have also been exposed to tales of other
world maps or even given copies of manuscripts from other places.
Why else would he have had the impetus to draw his own world map?
It's certainly reasonable to suggest that Fra Mauro knew about Islamic
maps in particular because of the many centuries of trading between
Venice, the Levant, and Byzantium. He also had at his disposal the
notes from Venetian world travelers such as Marco Polo and Niccolò
de' Conte (see also chapters 3 and 4).

If Fra Mauro knew all these maps or not, looking back provides the
context of world cartography in which Fra Mauro's map was set, and it
also prefaces the discussion about why his map, hung in 1460, was such
a game-changer. To understand that impact, we must know what went
before, which traditions were accepted, which were rejected, and what
the role of maps in their various cultures might have been.

THE VERY FIRST WORLD MAP

After the early, primitive maps discussed in chapter 1, around 600 BCE, mapmaking became a very different sort of endeavor. It branched out from scenes of local space and practical directions to whole fantasies about how the greater world might look and created the world map as an entity. Where and why did that change happen?

The oldest surviving world map comes from the so-called cradle of civilization around the Tigris and Euphrates Rivers in present-day southern Iraq. Inscribed on a clay tablet, this first world map was uncovered by archaeologist Hormuzd Rassamin in 1882. Unfortunately, Rassamin didn't grasp its significance as a world map simply because he couldn't translate the cuneiform, or wedge-shaped writing, gracing the top third of the tablet. [6] The tablet is now known as *The Babylonian Map of the World*, or *Imago Mundi*. [7] This first world map is a tiny bit of smashed-up clay, about the size of a human hand when glued together. It's dated between 500 and 700 BCE. [8] In photographs, this piece of flat clay appears to be a grayish color, but it's impossible to know if that coloration is original or if the clay changed over the centuries.

The reconstructed Babylonian tablet is composed of eight to ten pieces with a hole in the center, which presumably marks the center of the Babylonian Empire, perhaps the very city of Babel. The lower two thirds of the tablet is incised with two circles and five triangles sprouting from the outer edge of the circles like rays of the sun. There were clearly two other rays to complete the image, but they have been broken off. The writing on the surviving triangles marks these as areas full of uncivilized creatures, dark places outside the empire to be avoided, and if this tradition of inciting fear from monsters and barbarians started on the Babylonian clay tablet, it was certainly carried forward onto much later medieval Christian maps. [9] What appears to be a cross, with one rectangle placed in the center north to south and a shorter rectangle crossing east to west on its upper part, is actually a geographic representation. The vertical rectangle is the Euphrates River, and the horizontal rectangle represents the city of Babel. That cross stands on another horizontal rectangle covered with marks and this has been interpreted as marshland. Circling the rectangle are dots that indicate various cities or countries, such as Armenia and Habban. The concentric circles frame the land as the ocean, or salted water called the *Marratu*.

Clay tablets in general were the first permanent record of human behavior, the first pages of our history books because they archived daily life and how people accounted for their goods.[10] In a sense, these clay tablets are also the first account books because they often tracked items as they came and went. Some also contain text in cuneiform writing, which had been invented around 2,400 BCE. These written notes tell stories and myths that explain the thought processes of their makers and the belief systems of that culture. Because of the writing, these tables represent the first notes and books, and so they are the first evidence of written communication among people.

The Babylonian Map of the World represents a belief system because it is not about numbers but about place; it is not an accounting but a vision. The Babylonian Empire had gone through various iterations, but during the time this map was made, it was a powerful state and remained so until it was conquered by the Persians. Because of the high status of the Babylonian Empire at the time this first world map was made, we can assume the incised image was not just a picture of the empire but also some sort of political statement, or a symbol of Babylonian dominance and identity. What also makes this first world map interesting is how the maker put their own city in the center and then imagined a world that emanated from there. Surely this was a giant leap for humankind. That approach to mapmaking—centering the map by one's place—continues today as the standard for cartographers. *The Babylonian Map of the World* also prefaces the use of a round map and the structure of land surrounded by "salty water." Even more telling, the triangles coming off of Babylon and surrounding areas are points of warning about unknown lands and scary beasts that will eat you if you wander far from home, another trait seen in later maps.

This first world map, then, is itself a road map on how to draw a picture of the world and it established traditions in cartography that have been carried down the ages.

CLASSICAL GREEK AND ROMAN WORLD MAPS

The story of classic world maps is a complicated one, not only because these maps were drawn centuries ago, but also because so many have been lost or redrawn over time. In addition, the various texts that might have accompanied these ancient maps are secondhand. They have also been lost, found, rewritten, extrapolated from other documents, translated, and imagined from oral accounts.[11] As such, discussion of Greek and Roman maps usually relies on existing copies that were often made centuries later, or on reconstructing lost maps from various texts that have survived without their illustrations. The problem with the copies is that it's impossible to know what is left out or if the substitute cartographer was adding contemporary bits. In that sense, we are walking blindly, or at least visually impaired, through the history of world maps, reaching out to a very long-dead geographer and trying to decipher unreliable duplicates as we go.

Sometime between the sixth and seventh centuries BCE, a Greek scholar and astronomer named Anaximander proposed that the earth was floating, by itself, in space. For that, he is sometimes considered the father of astronomy and cosmology, even the father of scientific thought. But Anaximander was wrong about many other celestial issues. He thought the earth was a cylinder, not a sphere, for example. But his idea that the earth was not supported by anything opened the way for the possibility that planets came and went (that they orbited), and that the universe was always in motion. Anaximander also drew one of the first world maps, which in Greek are known as *oikoumene*. His map is lost, but others have reconstructed this map as a simple circle of water surrounding three continents.[12] The outer circumference of water represents the oceans. The land that makes up the center of the map is divided roughly into large tracks of areas labeled as Europa (to the north), Asen (Asia) to the east, and Libyen (Libya, which was another name for Africa, to the south). Water that we would call the Mediterranean, Adriatic, and Aegean Seas, seeps among these continents. There were also rivers marked on Anaximander's map, including the Nile. Greece, not surprisingly, is at the center.

Anaximander was a polymath of the highest degree, dabbling in astronomy, mathematics, and geography. It's too bad the original map does not exist, because the details would highlight the stage of geographic knowledge at that time. Instead, we must rely on the next Greek mapmaker, Hecataeus of Miletus (550–476 BCE), who made his own world map by copying Anaximander's map and filling it in. Hecataeus wrote two books on coastal surveys of Europe and Asia, and his world map was included in those surveys. Hecataeus called his two volumes a "world survey." This type of survey was called a "periplus," which was a logbook explaining coastlines and ports and giving the real distances between them. Such detailed directions suggest that Hecataeus might have been a sailor at one time, or maybe he had access to accurate information about coastlines from those who had traveled the seas. In a sense, these Greek peripli were like the much later portolan charts (see chapter 3), which also focused on coastlines and routes and plotted or explained directions for ships. But unlike later navigational charts, Hecataeus included descriptions of people and places pointing to his wide-ranging knowledge about other lands.

His world map is basically like Anaximander's but more cluttered. There are rivers, names of countries such as Egypt, Syria, India, and cities such as Syracusa in Sicily and Carthage in North Africa. The Caspian Sea sits in the northeast. We also owe to early Greeks the idea of using a grid to decide where geographic features might go. The Greek scholar Dicaerchus was the first, in about the early third century BCE, to use one horizontal and one vertical line to plot the earth. He supposedly drew a world map with that simple plot as part of a manuscript called *The Circuit of the Earth* (*Periodos gees*), which is, of course, lost.[13]

Eratosthenes, a Greek scholar and chief librarian of the Library of Alexandria around 230 BCE, is probably best known as the first person to measure the circumference of the earth using an ingenious method based on the understanding that Earth is a sphere.[14] He employed basic principles of geometry to measure the angle of the shadow of the sun in one place compared to its known distance to another place where the sun was directly overhead. Eratosthenes then multiplied that figure by the 360 degrees of a ball. He was surprisingly accurate—only off by ninety-nine miles at the equator. The content of ancient Greek world maps changed dramatically when Eratosthenes wrote his three-volume work *Geography* (*Geographika*) around 190 BCE. Those books earned him the title "Father of Geography." The third volume was all about a map of the world, but it is missing. The book might have disappeared in 48 BCE when the other two volumes perished in the great fire that destroyed the Library of Alexandria.[15]

We know about Eratosthenes's books because other Greek and Roman scholars had read them, and they passed on the knowledge. These scholars also reconstructed Eratosthenes's world map as a rectangle with longitude and latitude lines. This reconstruction echoes the former circular Greek maps because it, too, has three big continents and a huge Mediterranean Sea at the center.[16] But this map includes the islands of Britannia (England), the Arabian Peninsula, and India is greatly expanded. The Caspian Sea is now embedded in the northeast and the Persian Gulf lies to the south. Most importantly, Eratosthenes was the first to use latitude and longitude as a scaffolding to place land and water across the known world. His five lines of latitude were calculated by the length of days, and he considered them climate zones.[17] His map was also informed by the conquests of Alexander the Great, which had occurred two hundred years earlier.[18] That meant Eratosthenes was utilizing real-time observations of Alexander and translating them onto a map. Although such a resource seems obvious to us now, back then it was a leap to have someone at hand who had traveled outside the home country and bringing back reliable observations. This type of secondhand information would prove essential in later maps, especially the one made by Fra Mauro, which depended heavily on stories from travelers, traders, and explorers. In general, Eratosthenes's map is much more detailed than its predecessors' more diagrammatic forms; to the modern eye, it looks more like what we think a map should be—with grid lines, bumpy mountain ranges, enclosed seas, and open oceans. The continents are all out of proportion to our modern sensibilities, but this map looks more like reality than the conceived schematic fantasies that came before.

About two hundred years later, the Greek scholar Strabo published a seventeen-volume work called *Geographica* (first published in 7 BCE and we have an extant version) in which he waxed lyrical about lands, sea, people, and cultures. That work was influenced by Greek exploration and also by the conquests of Alexander the Great.[19] Although Strabo did not draw a map, others constructed an approximation based on descriptions in those volumes, and that map is much like Eratosthenes's.[20] At the same time, Strabo did not believe that people could live in the far north because of the cold, and he suggested a very short Africa ending at the latitude of Ethiopia.[21] He also shortened the world from east to west.[22]

Marinus of Tyre, another Greek geographer, imposed latitude and longitude on world maps in the first century CE, and for the first time, China appeared on a Western map. We only know of Tyre's contribution to cartography because the greatest and most influential Greek geographer of all time, Claudius Ptolemy, had read and quoted Tyre's work in his eight-volume masterpiece *Geography* (*Geographike Hyphegesis*), which is considered by some merely a revision of Tyre's work.[23] Ptolemy was an astronomer, mathematician, and astrologer living in Alexandria, Egypt, and he wrote many manuscripts and books on those subjects. *Geography* was written in 150 CE, and it is, in many sections, a manual for making maps. Ptolemy even defines geography as "a graphic representation of the whole known part of the world, along with things occurring in it."[24] The books provide a catalog of more than 8,000 geographic features and places and 6,300 of those pinpoints have corresponding coordinates of latitude based on day length and longitude based on the number of hours a place is to the east and west of the city of Alexandria, his prime meridian.

Ptolemy also famously tackled the issue of projection—that is, the distortion that comes from trying to take the skin of a sphere and accurately spread it out onto a flat surface of paper, parchment, wood, or animal skin and keep everything in proportion.[25] Projection of the earth's surface is a challenge that world cartographers continue to wrestle with (see chapter 6). It's simply not possible to take the covering of a ball, smooth it out, and keep the shapes of continents and seas in perfect proportion to each other. For example, hold an orange and draw some figures on it with a marker. Then peel that orange, navel to bottom, into strips that are of equal size at the orange's equator. Then flatten those strips. The bits of orange peel will be long with pointy ends and wide middles. Placing them next to each other produces a zigzag sort of object that is only closely connected at the middle circumference. Both the top and bottom ends of the strips are long triangles with lots of space in between. Cartographers always want a way to join those strips and create a continuous flat map of the world, and that goal has haunted them forever.

Tackling this problem, Ptolemy began with the simplest plan based on the idea of longitude and latitude as a working grid. For his first projection, Ptolemy imagined those horizontal and vertical lines lifted off the ball and placed on a flat surface. The lines of longitude would start close to each other at the top (the North Pole) and then fan out east and west. At the equator, the fattest part of the ball, those lines should drop down as if going over a cliff. In this design, the horizontal longitude lines are curved, and the scheme looks like a cone. Thus set, the lines produced a graticule, an overlaying grid system of coordinates that give spatial conformity to the earth's globe.[26] Ptolemy's second projection worked to take into account the curvature of the earth both north to south and east to west. Called the cloak projection, both the lines of longitude and latitude are curved.[27] Ptolemy thus gave the world a reasonable flat field for plotting and placing known geographic features.[28]

Surprisingly, Ptolemy didn't attempt to put his ideas into practice and never made his own map. *Geography* was lost, but then reappeared in Arabic in the ninth century and showed up in the West in the twelfth or thirteenth century, but it didn't gain a large audience in the West until it was translated into Latin in 1406.[29] Ptolemy's *Geography* then had a tenacious hold as the only cartographic instruction book for a thousand years thereafter. But while the Latin version of Ptolemy was greatly influencing Western mappamundi cartographers, and they had a hard time letting go of Ptolemy's ideas, their knowledge of the world was expanding exponentially because of global exploration and trade. And yet, cartographers of the Middle Ages and the Renaissance clung to Ptolemy's dictates, even though he was inaccurate in many ways and his projection produced a world map with many continents out of proportion to reality. For example, in 1467 (ten years after Fra Mauro's map) Nicolaus Germanus rewrote the Latin version of Ptolemy's *Geography* and was the first to draw a world map based on the second projection. He also added Scandinavia to a map for the first time. More to the point, this map was made according to Ptolemy's directions 1,200 years after the great Greek mathematician had come up with the system for projections.

ISLAMIC MAPS

As the Turkish scholar Fuat Suzgin has maintained in his seventeen-volume work about the Muslim contribution to the birth and growth of science, *Geschichte des arabischen Schrifttums*,[30] we need only turn to the Muslim world to see how cartography was progressing from 800 CE to the Middle Ages and beyond. These scholars contrasted with cartographers in the West because they were not just culturally and religiously in another sphere but physically distanced as well. Their world, and knowledge of the world, was centered on the Middle East rather than the Mediterranean. That alone made for maps that contrasted with those from the West or the Far East. Also, Islamic scholars were superior at mathematics and astronomy, both essential tools for mapmaking. Arabic scholars and geographers were also ahead of those in the West because they had already translated Ptolemy's instructions in the ninth century, and they were open to information from the West, including Greek geographers and philosophers. And so, early on, Islamic geographers had a good road map for figuring out projection and placing continents. [31]

At the same time, Islamic maps of the world look different from Christianity-based maps from Europe because these mapmakers were not interested in portraying the real world in minute realistic detail. At first glance, to some, Islamic mappamundi may seem simplistic, and they might be set aside because they don't chart the exact shape of coastlines and such. Scholar Karen Pinto explains, however, that these particular cartographers were choosing to be "carto-ideographs," meaning they decided to use a more stylized representation of the world, smoothing out the routes and connections to make them more easily read and followed. [32]

These maps are also often pieces of grand art and intended as such. They were most often produced for the elite, and the paints contained ground-up precious metals and gems. Islamic cartographers were also, according to Pinto, carrying cultural memes that were passed along by viewers and owners of the maps. In other words, they were cultural documents, not road maps for getting somewhere, nor documents about where to go to conquer other lands. Instead, these maps were created on purpose as carriers of cultural knowledge and traditions. They simply didn't need all the squiggles and features of Western maps.

For example, the great Islamic scholar and algebraist Muhammad ibn Mūsā al-Khwārizmī composed his explanation of the earth, also called *Geography* (*Kitāb Ṣūrat al-Arḍ* or *Book of the Depiction of the Earth* or *The Image of Earth*) in 833 CE. [33] He included over 2,400 coordinates on the map based on calculations of longitude and latitude. [34] This work also initiated a long-lasting discussion among Islamic cartographers about projection and the mathematics of determining coordinates. [35] For example, al-Khwārizmī corrected many errors of Ptolemy, such as his exaggeration of the Mediterranean Sea. He also clarified the size and position of areas of Islam, and much of Africa and the Far East, areas all less known by Western cartographers. In addition, al-Khwārizmī's calculations made sure that the Atlantic and Indian Oceans were open bodies of water; Ptolemy thought they were landlocked. The original manuscript probably included a map, but it was lost. Following earlier traditions in mapmaking, al-Khwārizmī's world map was later reconstructed from his writings in 1972.[36]

Abu Zaid Ahmed ibn Sahl al-Balkhī (850–934 CE), born in Afghanistan, was a scholar who studied everything from poetry to mathematics. He also prefigured modern mental health care when he made a distinction between psychosis and neurosis, made categories of various mental disorders such as depression and anxiety, and initiated what we now call cognitive behavioral therapy. One of al-Balkhī's other significant contributions was a geography book titled *Figures of the Regions* (*Suwar al-aqalim*), which established the Balkhī school of terrestrial mapping. Established in Bagdad by Abu Zayd al-Balkhī, this approach was contrary to Ptolemy's instructions.[37] Unfortunately, none of the maps from that school exists today, so it's hard to judge exactly how the new approach might have changed cartography. And yet a re-creation of the al-Balkhī school of world map show a symbolic sort of design with straight or slightly curving dark thick lines for rivers parsing up the land. This map, with its bold graphic simplicity, seems to signal the birth of so many maps made today, where geography is smoothed out into thick lines made simple for convenience, such as subway maps. These are also the first maps to be oriented to the south, the same direction used by Fra Mauro 450 years later, and he was a Catholic monk. It's impossible to say exactly why Fra Mauro made that decision, but since he was presumably not following the dictates and practice of Islam or Islamic culture, it might just have been acknowledging the contributions of Islamic cartographers along with trying to break from Christian faith-based maps.

The great Islamic traveler, historian, geographer, and student of human cultures al-Mas'ūdī (893–956 CE) aided these eighth- and ninth-century mapmakers. He was born in Bagdad, but as a young adult set out the see the world. At that time, the Islamic empire was vast, but he went even further. Al-Mas'ūdī traveled to India, China, Sri Lanka, Indonesia, Malaysia, Madagascar, and the coast of East Africa. Around 922 CE he wrote his first book about his travels, titled *Meadows of Gold and Mines of Precious Stones* (*Muruj-al-Zahab wa al-Ma-adin al-Jawahir*), and that book contained a world map. Al-Mas'ūdī was the first to base his map on his own experiences combined with information from others, and he was the first to eschew religious, or spiritual, overtones on a map. The al-Mas'ūdī map, as reconstructed, is a circle oriented south and has the usual large landmasses of Europe, Asia, and Africa surrounded by a ring of water. But he also has another landmass in the southwest (if the map is flipped around) under Africa. Al-Mas'ūdī had written about Arab sailors who in 899 had gone west, found land and "natives" there, traded with them, and came home. It's no great leap to realize this adventure might have been across the Atlantic to some part of North or South America, or the Caribbean.[38]

In the eleventh century, Islamic polymath Abu Rayhan Muhammad ibn Ahmad al-Bīrūnī calculated the circumference of the earth and thereby suggested there had to be another landmass to the west of Europe on the way to Asia. He was certainly right about that. And this was hundreds of years before Christopher Columbus. Al-Bīrūnī also drew a schematic world map that was later copied, or replicated, by others such as al-Qzwīnī (Zakriya al-Qazwini).[39] This map of the world is a simple circle drawn as a red line with a smaller half circle inside representing landmasses. Four rectangular fingers are rising from the top of the half circle and there are two indentations at the bottom of the inside half circle. Flipped around and oriented north, this is obviously a map of three continents attached to each other with indentations representing the Mediterranean, the Red Sea, the Persian Gulf, and the Indian "Sea." A circle in Asia is probably the Black or Caspian Sea. The map is basically a sketch, an outline with labels, and that's all. The beauty is in its simplicity and the fact that it directly and clearly communicates the known world.

Around 1020–1050 CE, some Islamic cartographer wrote and drew maps for a manuscript on paper titled *The Book of Curiosities of the Sciences and Marvels for the Eyes* (*Kitāb Gharāʾib al-funūn wa-mulaḥ al-uyūn*, which rhymes in spoken Arabic). It was then copied sometime in the early thirteenth century.[40] Today we only have copies of this work set out in two volumes, one about the heavens and the other about the earth.[41] These books contain seventeen maps or cartographic sorts of designs that set the bar for the time. The second volume, which is about the earth, is made up of twenty-five chapters, some of which are missing pages.[42] Scholars have determined that both books were probably composed in Egypt. One of the extant copies was acquired by the Bodleian Library at Oxford University in 2002, and it contains a rectangular world map and a circular world map. The round map has the usual three continents surrounded by water and an Africa that takes up much of the southern section of the world. The Indian Ocean is landlocked as a sea. This map also has an overlaying grid of climes or horizontal lines that represent the length of the day, which, as in old Greek maps, were of latitude.[43] The rectangular world map in *The Book of Curiosities* has received more attention because it is so unusual. It is the first world map to have a graphic scale on the edge, which was presumably an attempt at using mathematics to place geography more accurately.[44] In the middle of that scale is a drawing of a half circle representing the Mountains of the Moon, with five rivers branching off, suggesting the mouth of the Nile River. The landmasses are composed of three continents with many projecting fat fingers of land, and the coasts are smoothed out with no indentations or frills as in real coastlines. There are no islands in the seas or oceans, but they appear in other regional maps in the book, and so the world map is intended as more schematic than detailed. Place names are in Arabic, and areas and cities important to Muslim culture such as Mecca are highlighted. This might also be the first presentation on a map of the evil kingdom of Gog and Magog. They were two monsters or barbarians who lived outside the known world and were considered evil, threatening beasts with followers who did their bidding. But there is no evidence that

Gog and Magog were real creatures or that they lead real kingdoms full of savages ready to overrun the rest of the world. Instead, time and further exploration of the world proved they were just mythical characters, but they appeared over and over on later Christian maps. On this map, Gog and Magog were kept in check with a gate supposedly put there by Alexander the Great. *The Book of Curiosities* world map also has several lines of red dots that cross the world in places here and there. These dots obviously indicate travel routes. Those various routes stand out visually because the cartographer put his world through a house of mirrors and enlarged some parts because they were more meaningful for Islamic trade and life while shrinking others, like a modern-day equal areas map (see chapter 7). This map of the world seems to focus on the commercial business of the Muslim world in the Mediterranean, especially the Arab trading triangle of Sicily, Egypt, and Tunisia.[45] According to scholars, the cartographer was more interested in history and descriptions of places and people, rather than the Ptolemaic mathematics at its core. The extant map has obvious flaws, which are probably the result of an incompetent copyist who also wrote incorrect Arabic text.[46] And some of the trails of red dots are not even identified. More interesting, there is a bit of land labeled *Inqiltarah* (Angle-Terre), which would be the first time Great Britain received the name "England" (in Italian, Britain is still known as Inghilterra).[47] Unfortunately, the circular map is much like other circular maps that came before and later, so it has received less speculation from scholars than the more unusual rectangular map.[48]

Perhaps the most significant and well-known Muslim cartographer of the Middle Ages was al-Sharīf al-Idrīsī. His family came from North Africa, but his father settled in what is now Spain. Al-Idrīsī was born there in 1100 CE and traveled extensively as a teenager and adult through both areas and the rest of Europe. He ended up in Sicily, in the court of Norman King Roger II, where he worked for eighteen years. In 1154, with the collaboration of twelve other cartographic scholars in the king's court, al-Idrīsī completed a circular world map.[49] His cartographic approach was surprisingly aligned with Ptolemy rather than with his Muslim predecessor in geography, Ibn Hawqal, who represented the Muslim view of the world according to al-Balkhī.[50] Then, after fifteen years of discussion with other scholars and collating all the information, al-Idrīsī and his collaborators decided to draw all the information on one large and permanent map. They drew it first on some usual surface, such as velum or wood, and then copied that drawing on a huge disc of solid silver, and this map incorporated latitude and longitude. Al-Idrīsī was also instructed by King Roger II to make the map to scale, and that it be graphically accurate. In other words, the king wanted to know what his world, and the world in general, looked like.

To accompany the silver world map, al-Idrīsī complied a supportive geography written in Arabic and Latin titled *The Pleasure Excursion of One Who Is Eager to Traverse the Regions of the World* (*Kitāb nuzhat al-mushtāq fī ikhtirāq al-āfāq*), also sometimes called *The Book of Roger*, since the king paid for it.[51] This book also contained one small world map and seventy-three regional maps, all following the dictates of Ptolemy, including dividing the world into seven climate zones. The regional maps were also taped together to form a rectangular view of the world. Scholars have suggested that the section maps in al-Idrīsī's travel book are probably based on the silver world map rather than the other way around.[52] Although the silver disc and the original book manuscript no longer exist, al-Idrīsī had made a second book for King Roger II's son, William II, and that book was reproduced in manuscript form centuries later, with and without maps. Those copies are held in various libraries such as the Bodleian Library at Oxford University. Presumably, the small world map in the book was an echo of the silver disc since the book and the larger map are associated in their time of creation, and the cartographers and patron were the same for both. The world map from the book is also round, with a ring of deep blue as its circumference outlined by a ring of flames. Most notably, the map is oriented to the south, meaning that it is upside-down. That orientation had become usual in Islamic maps because Mecca was south of where most believers lived and so they faced south for daily prayer. This was a faith-based decision, much as later Christian maps faced east, toward Jerusalem.

There are seven evenly spaced red horizontal lines curving across al-Idrīsī's Earth to denote the seven climates (*klimata*) from tropic to temperate, but these marks also function as lines of latitude. This world map has the same spread of continents as earlier Islamic maps with Europe, Africa, and Asia. Africa is especially huge and mostly blank, but Europe, where al-Idrīsī had traveled and where he lived, is more detailed with mountains and place names. The Mediterranean appears much larger than it is, and Sicily is oversized as a nod to King Roger II and his kingdom. Blank blobs of islands also dot the entrance to the Mediterranean and then north and south along the Atlantic coast. They also fill the areas we now know as the Indian Ocean and the Caspian Sea. Historians think these dots represent known islands such as the Canaries, but they are also used to denote unknown areas, such as a passage through the Indian Ocean or up and down the Atlantic. Cartographic historian Jerry Brotton points out that while Christian cartographers stuck to the Bible when centering their maps and placing the Garden of Eden on the map as if it were a real place, Islamic cartographers were less likely to refer to the Koran's description of paradise. They seemed more interested in displaying economic and political issues on their maps by including trade routes and the various administrators of cities and countries.[53]

In the early 1200s, the Persian cartographer Zakarīyā Ibn Muḥammad al-Qazwīnī, famous for his cosmography *The Wonders of Creation* (which is still available today), wrote *Monuments of the Lands* (*Athar al-bilad*).[54] Both books have maps, including a circular map of the world, oriented south. He closes off the Indian Ocean with Africa, as Ptolemy did, and his Atlantic Ocean dominates the map with scalloped blue lines. The al-Qazwīnī map is also decorated in gold and silver. This map is more symbolic than informative; it has few inscriptions as in the style of al-Bīrūnī, but it was part of a large written manuscript, so the map functions as an additional illustration.

Islamic cartography seemed to wane in the thirteenth century, but in 1349, Ibn al-Wardī wrote *The Pearl of Wonders and the Uniqueness of Strange Things* (*Kharīdat al-ʿAjāʾib wa farīdat al-ghaʾrāib*).[55] This book covered geography, climes, animals, plants, government, and people around the known Arabic world. Al-Wardī also included a map, which was a typical circle with a fluted edge which had been used before. Like much older Islamic maps, there is more land than water, and it's graphically simple, but his placement of the Nile was spot on. Again, this map is accompanied by text and so it need not be covered with inscriptions, but al-Wardī did include mountain ranges and place names.

Islamic cartographers gave the history of mappamundi both the southern orientation and the possibility of symbolic and artistically beautiful maps. Although Fra Mauro chose his map to be more complex in terms of text than these maps, he did follow their lead in making a beautiful piece of art, and in his map's orientation to the south. Looking at Fra Mauro's map now, hints of Islamic cartographers peek out through the many undulating lines that denote waves, and in his choice of color palette that delights the eye.

ASIAN MAPS

Although the Chinese and other Asian peoples were adept mapmakers of their cities and regions, there are few Asian world maps from antiquity. China, after all, is a vast landscape, which might mean it was unnecessary to look much farther than outlying territories to imagine the whole world. Or perhaps Asian countries then were not as dependent on seagoing international trade as countries to the West, where travel to other countries and cultures was an integral part of life, and so they need not figure out trade routes or other cultures which they rarely encountered.

Although the Chinese have been drawing maps since the fifth century, the first of what might be called a world map from the Chinese perspective was created in 801 CE by scholar Jia Dan (also called *Dunshi*). He also wrote a forty-volume work of geography that explored territories of China and included areas they had lost on his Map of Chinese and non-Chinese Territories in the World (Hainei Huayi Tu). Jia Dan's world map was 30 by 33 feet (9.14 x 10.05 M) with a grid system for scale. This map was lost, but some of the information later appeared on another map cut on steel in 1136. The best-known world map from China also comes later in time, in 1390, and it was painted in color on silk, about 12.5 by 14.5 feet (3.8 x 4.42 M), and called the *Da Ming Hunyi Tu*, or world map. Of course, China is at the center of this world, with Java, Japan, Mongolia, Africa, and Europe as surrounding places.

An Asian map of the world was made about sixty-five years later, in Korea, around 1402. Called the *Map of Integrated Lands and Regions of Historical Countries and Capitals* (*Honil Gangni Yeokdae Gukdo Ji Do*), or more simply the Kangnido map, it was drawn by cartographers Yi Hoe and Kwon Kun.[56] This, too, is a Chinese-centered map, and it was made with reference to precious Islamic world maps. It therefore includes Europe and Africa, and interestingly it shows that Africa can be circumnavigated.[57] Although this map was made before Fra Mauro started his work, it's doubtful that the Venetian monk would have known of its existence. Two copies survive today—one on silk, and the other, much larger, painted on paper. There is certainly a long and complex history of many other maps from Asian and Southeast Asian societies, but these maps are regional, or even more specific.[58]

EARLY EUROPEAN WORLD MAPS
LONG BEFORE FRA MAURO

In terms of the history of cartography, Europe was ground zero for the belief in, and promulgation of, Christianity, and this is the main ambiance in which Fra Mauro made his map. Because of the push for illustrating Christian dogma, European maps from various countries or territories have common features and symbols that connect them in a religious way fundamental to Christianity. Those features include the Garden of Eden, Earthly paradise, Jerusalem as the place where Christianity was born, the Crucifixion of Jesus, angels that appear as real beings, and so on. Chapter 4 explains what those features are, how Fra Mauro distinguished his map by altering, eliminating, or repositioning those items, and how those changes moved his map more toward science than religion. But other maps from the seventh century onward made in Europe were dedicated to presenting a Christian way of life, and they were used as political documents in that way as well. For example, the Garden of Eden is commonplace on European maps and yet we know there is no real Garden of Eden anywhere on Earth.[59]

Europeans named their maps of the world "mappamundi" (or mappa mundi) in the late twelfth century as geographers fiddled with the scale and organization of the way they presented their view of the world. Most European mappamundi were not necessarily intended to be geographically accurate representations of the world because their job was to tell the story of human history through the Christian lens, and so they are all biased and skewed.[60] That kind of cartographic bias continues today, but it is not based on religious belief but on political or geographical identity. For example, we assume that contemporary maps are all accurate, especially given the technology we now use to see the world from afar, but we still have a major issue with orientation. Some contemporary world maps place the United States in the center, while others put Europe or Russia as the focus, which reinforces the idea that one country or continent is the center of the world. World maps with Australia as the center are unsettling to everyone but Australians. The center of Fra Mauro's map is, unusually, the Near East, or what would be present-day Iran. Italy, the friar's home county, which is easily noted on any map because of its boot shape, is at the east (or west if you reorient this map to the north). In other words, Fra Mauro chose not to center the Christian capital, Jerusalem, nor his homeland, which is an interesting, and significant, break from the map traditions of his time.

The story of European world maps begins in Spain. Isidore, the Archbishop of Seville, Spain, composed an extensive twenty-volume encyclopedia titled *Entymologia* in the early 600s. This set of books was a compendium of all classical knowledge up to that time. Isidore might also be the inventor of the first form of medieval map called the T-O map (or T and O, short for *orbis terrarium*) to represent the world, and it is a very ancient schematic.[61] Although Isidore did not draw this form himself, it is implied in his encyclopedia where he wrote that the earth is made up of three landmasses: Europe, Asia, and Africa. In the T-O scheme, a big sea (the Mediterranean) divides Europe from Africa, and the Nile River separates Africa from Asia. Orienting that idea to the east puts Asia at the top with a *T* of water underneath, which separates Asia from Africa, and Europe further separates those continents right and left. In a T-O map, the ocean surrounds the circle, forming the *O*. Like so many maps after Isadore, the T-O form is a simplified graphic, a representation of landmasses separated by water. No one would use such a map for directions; it's not so much a map of the world as a logo of the world.

Isidore also influenced later cartographers of Spain such as Beatus, a Spanish monk who followed Isidore's conception of the T-O world but drew his late eighth-century maps as a round-edged rectangle rather than a circle, perhaps to accommodate a manuscript or codex.[62] The Beatus map is divided into three continents, surrounded by ocean as in that basic T-O structure. But this early world map is much more complex than the design suggested by Isidore. It includes strikingly bold blue marks for rivers and seas, and it has place names. The point of this map was not to educate about the shape of the known world, but to show how the apostles had spread across the world carrying Christianity with them.[63] The Beatus map is also an evangelical track record of spreading the faith after the fabled Christian Apocalypse. This fable, or vision, is laid out in the New Testament whereby the world is destroyed, heaven is then at war, and Jesus Christ comes back triumphant.

Another world map from the second half of the eighth century is held at a library in Albi, France. This map is drawn on goatskin and is very tiny for a world map, only 10.5 by 8.5 inches (27 x 22.5 cm). The Albi map also stands out in the history of old maps because its world is *U*-shaped. The map is oriented east, and the center of the horseshoe is the Mediterranean, which is painted grayish blue. Europe, Asia, and Africa blend into each other as one big landmass surrounding the Mediterranean. In that sense, this world map is rather primitive, with few inscriptions and only five identified islands in the Mediterranean, but then it is quite old.

There are no known world maps from Europe for the next three hundred years, but since maps were drawn on all sorts of perishable materials, the lack of examples is probably more due to destruction and loss than an absence of mapmakers. It would take a set of circumstances, and a certain British personality, to rescue and value early maps and try and fill in the gaps.

Sir Robert Cotton was a British gentleman and bibliophile who helped save innumerable ancient manuscripts in the early sixteenth century when King Henry the VIII started closing monasteries, abbeys, and convents and selling off their contents to line the royal coffers. Cotton's grandson later donated his vast collection to Great Britain, and it became the basis of the British Library. Among the manuscripts saved by Cotton was a world map, called the Anglo-Saxon Cotton map, which is dated around 1040 CE. No one knows who drew this map, or what it was based on, but it is a world that looks familiar to those familiar with medieval maps. The Cotton map is rectangular, oriented to the east, the continents of Asia and Africa, if that's what they are, are connected. The seas and oceans, all dotted with islands, are gray, and the land is off-white. The Red Sea and the Persian Gulf, as well as large rivers in Africa and Asia, are bright red. The land is not symbolically executed as in other maps of the previous period but waving at the edges like real coasts. There is no Garden of Eden and Jerusalem is not the central point of the map. In that execution, this medieval map does not follow the Christian dictates like other maps of the time, but some sites pinpoint biblical events, such as where Noah and his ark landed, and there are some explanations of imagined creatures. The main inscriptions are in Latin or Old English. Most notable are the details of Great Britain, including Scotland and Ireland, and various British islands such as Orkney, the Isles of

Wight and Man, and the Channel Islands. This kind of British detail is not found in other world maps of the time, suggesting it was made by someone from that area.

The British also made the Sawley map around 1190. It was named after an abbey in Yorkshire and is a religious production.[64] This map is part of a larger manuscript; the map sits on the left-hand first page of that manuscript, as if it were the preface or endpaper for a text on geography, astronomy, and history. But no one is sure if this map was just placed there or if it was part of the original manuscript. Notably, it is very small— 8.5 by 11 inches (9.5 x 20.5 cm)—which is just right for manuscript inclusion. The map is more ovoid than round, and it has the usual green ocean surrounding the world while purple rivers snake through the landscape. There are 229 inscriptions in legible Gothic script across the landscape. It also has miniature buildings and what appear to be fences in several places. For the first time on a Western map, the scary savages Gog and Magog are there on the upper left. An angel is pointing right at Gog and Magog, who apparently live on islands off the coast of the world just waiting for Judgment Day so they can take over. The Sawley map is, as expected given its biblical nature, oriented to the east, although Jerusalem is not in the center. Perhaps the most striking feature of the Sawley map is the four fully winged angels in each corner all pointing toward the map. The Sawley map is also a Christian guidebook. It walks the viewer from the Garden of Eden at the top, down through where the Tower of Babel was supposed to be, and across the Holy Land.[65] The Sawley world comes across as a clean and simple world, but a busy and religious one.

EUROPEAN MAPS RELEVANT TO FRA MAURO

It is easy to make the case that Fra Mauro must have been aware of maps made in the 1300s and 1400s, especially those made in Europe. He probably had no primary access to any of those maps, but he would have heard about them. Fra Mauro might also have had access to copies, maybe drawings in library manuscripts. For all we know, there were copies of such maps in the library of the monastery of San Michele where he lived. But since Fra Mauro's notes are now lost (see chapter 4), we have no concrete evidence of what he knew about his predecessors. But the mere fact that he drew a complex and well-informed map shows that he was encountering older maps, drawing information from them, and making adjustments. In the inscriptions on his map, Fra Mauro also endlessly explained his choices, referring to common knowledge or specific ideas of others, and in doing so he was acting like any scientist who builds a case based on the work of others (see chapter 5). Although maps are not text manuscripts per se, they are resources, and surely Fra Mauro knew his resources. How else could have drawn his map?

Perhaps the most renowned Western mappamundi of the 1300s, the *Hereford Mappa Mundi*, was drawn in Britain, and it is still hanging in Hereford Cathedral.[66] It was first mounted on a side wall, then placed under the floor for protection, and finally unearthed and restored in 1855 by the British Museum. It was also protected during World War II as an important historical artifact. This map is huge—4.33 feet (132 cm) in diameter, which is enormous for a world map of that time, and it was designed to be hung on a wall. The map is signed by Richard of Haldingham and Lafford but must surely have been drawn and inscribed by a team with Haldingham in charge. It was probably made in the city of Lincoln, because that city is drawn in detail.

The map is on calfskin with the neck at the top, and there is a slight crease right down the center, which is the former spine of the calf. The Hereford world is round, and oriented east, with paradise at the center top. Jesus Christ stands above paradise, outside the map on the neck of the velum. Jerusalem is a dark-geared wheel in the center with a drawing of the Crucifixion above that city. This is the moment where Western maps began to put Jerusalem as the center of the world according to Christian dogma, and that collective understanding was probably underscored by the Crusades as the crusaders moved to "liberate" Jerusalem from the "infidels"—that is, Muslim control.[67] Also, the city had been sacked in 1244 and although it had been in Christian hands after the Sixth Crusade, it was lost again. That loss, map historian Evelyn Edson feels, turned Jerusalem into the status of a mythical city for Christians, as well as one with religious overtones, and made Jerusalem the center of the world for cartographers.[68] The map is also covered with Christian historical inscriptions that follow the Old Testament from the time of Noah's Ark to the life of Christ, and it portrays the myth that Christ will come again after the Apocalypse ends the world.[69] The map also has wild animals and human monsters at the edges of the world; this pattern of representing the unknown as inhabited by threatening creatures includes not only Gog and Magog but also a long list of mythical beasts and oddly shaped humans,[70] reinforcing the idea that what we can't see and what we don't know, that vast unkown, is scary and threatening.[71]

The Hereford map is nicely annotated in Latin with rivers and place names that are familiar today, and it is busy with tiny buildings, rivers, islands, and places that together work at "incorporating history, geography, botany, zoology, ethnology, and theology into one harmonious and dazzling whole."[72] Oddly, Augustus Caesar also appears here, with a mandate for those during Julius Caesar's reign to map the world back in 44 BCE; this is either a historical note or perhaps portrayed as a mandate for the ages.[73] The Hereford map seems to be a "God's-eye view" from above where the viewer can take in the whole world in one glance, and be forewarned about the coming Apocalypse.[74] Here we see the combination of geography and Christian spirituality, or, as map historian Evelyn Edson puts it, "The theater of human history in a union created by God."[75] All this is both overwhelming and frightening to our modern sensibilities but was presumably a reasonable viewing experience for medieval Christians.[76]

Another mid-thirteenth-century German mappamundi, designed by cartographer Gervase of Ebstorf, or perhaps someone else, was "discovered" in the nineteenth century in a convent.[77] It could hardly have been lost, given that this map measured 12 by 12 feet (3.6 x 3.6 M). It is a round map drawn on thirty goatskins that were sewn together. The world is, of course, oriented to the east, and centered on Jerusalem. At the top is the head of Jesus Christ, while his hands mark the east and west positions on either side of the circle and his feet are dipping into the ocean at the bottom.[78] In other words, this is the world, and it all belongs to God. The overarching presence of Jesus also speaks to his resurrection and the possibility of the "redemption" of the human race.[79] But the text is not just based on the Bible. It also contains nonreligious and pagan beliefs. There is writing all over this map, including descriptions of animals, definitions of things and places, and words all around the outside circumference of the world. Overall, this map is a fine encyclopedia of the European medieval world and how cartographers, and people at large, viewed their environment as a mixture of religion and suspicion. It is also a very busy map, crawling with places and people, and in that sense, it is a precursor to Fra Mauro's map, but it lacks the sparkling colors of the Venetian product. Instead, the Ebstorf map is a palate of beige and gray with some red and orange for accents. Sadly, the original was bombed to bits during World War II when the Allies hit Hanover and all we have left are photographs, mostly in black and white. But there is a much smaller anonymous version, a copy that is, of this map (3.74 by 1.45 inches or 9.5 x 3.7 cm), which was drawn in Germany in about 1260. Known as the *Psalter World Map*, it is book-sized and was probably part of a religious prayer book. That tiny world map is now preserved in the British Library.

The most significant German world map was made by the Benedictine monk Andreas Walsperger in 1448 just before Fra Mauro and his team set out to make a mappamundi. It is a round map of the world on a piece of parchment measuring 22 by 30 inches (55.7 x 75 cm). This is a visually appealing map with its green water, the Red Sea is bright red, and coastlines are not smooth but undulating in and out, although not in the extraordinary way they do on the Fra Mauro map. The extensive text in Latin codes Christian and Islamic cities with different colors—red for Christian and black for Islam. This map is also oriented to the south, like Fra Mauro's map, and surrounded by nine cosmological rings with floating heavenly spheres imposed across some of the rings. There are also half circles on one ring that denote various winds. The continent of Africa crawls across the water that surrounds this earth and pokes into the outer border, making it clear that there was no way around the southern tip. It has the usual Christian messages, Gog and Magog and cannibals are there, and the south is an uninhabited place full of monsters.[80] A large and lovely castle representing paradise sits in China with the four rivers that water the world, as expected of paradise, draining from the castle. Jerusalem is at the center, as per Christian tradition.[81]

Take away all the religious allegory and this map has striking similarities with Fra Mauro's map. It's as if the Walsperger map is a simpler, less cluttered, imitation of Fra Mauro's map, or a first, and less successful, attempt at an accurate word. It has a similar adherence to real coastlines, exuberant text, and such. But there is no way of knowing if these two cartographers, Mauro and Walsperger, knew of each other—probably not. The similarities in their cartographic styles might just be a result of the fact they were working at about the same time, with the same cartographic history, and both had been informed by the increasing movement of Europeans from place to place as the Age of Discovery began, which in turn led to knowing what far-away lands really looked like.[82]

PORTOLAN CHARTS AND THEIR
INFLUENCE ON WORLD MAPS

During medieval times, when world maps were becoming both fashionable as art objects and informative about world geography, there was another kind of map, called a portolan (or marine) chart, that had come into common use in the thirteenth century.[83] These maps were startlingly accurate, and they changed the course of cartography.[84] Portolan charts are navigational charts drawn and used by sailors, and so they are limited in their scope and detail while world maps were full of all sorts of information.[85] With portolan charts, ships could get from one place to another; world maps made it clear that there were potentially many places to go. Fra Mauro certainly knew about portolan charts, because he made at least one, which we'll explore in chapter 4. Also, as a Venetian, he would naturally know about any technology used for water navigation.

Portolans are based on pilot books, known as portolani, which are the running commentary written by pilots while sailing; many extant portolan charts accompany a pilot's book. They were usually drawn on sheepskin and most often oriented to the north, which is reasonable because the decision about orientation was practical, not theological. [86] Since the goal was to show routes used by sailors, these charts stand out visually because they focus on coastlines and pretty much ignore the interiors of countries and continents. Even the names of cities and other places are pinned along the coastlines; they stick out from the edge of land, indicating that the city is right there on the shore, or perhaps a little inland. This undulating strip of place names adds a frilly tex-tualized edge to the running coastlines of portolan charts. Since the names follow the coasts, the viewer needs to turn to the map to read them, which underscores the fact that portolans were designed to take aboard ships and used for ongoing navigation; they were not meant to be framed and hung on a wall. Also, portolans name only the mouth of rivers and there is nothing about a river's inland course because merchant traders were not the least bit interested in going inland. The rest of the various land masses are virtually empty. It's also easy to spot a portolan chart, as opposed to a map of the world, because they cover a restricted bit of territory; many of them concentrate on the Mediterranean and the Black Sea, where so much trading occurred among countries and empires. [87] There are also Islamic portolans that cover ports in the East that were important for Islamic traders.

Portolans are also immediately recognizable, as opposed to maps of roads, districts, or anything else, because of the many straight lines called rhumb, or compass rose lines, that cover these maps. These lines sometimes start at compass roses, intersect with them, or are pinpointed equidistant around the map's edge. From that anchoring point, the lines proceed straightforwardly across water and land. This technique allowed a pilot to note his starting place, find his destination, and then lay a straight line between them. He could then see the closest rhumb line next to the ruler and use that to calculate both his current bearings and the route to the destination. Unlike so many mappamundi, they were based on measurement, scale, and function. In that sense, portolans could be classified as the first maps based on reality and science.[88]

The oldest surviving portolan chart is Italian, from Pisa or Genoa made sometime time between 1275 and 1300 and it's called the *Carte Pisane*, after the city of Pisa.[89] The *Carte Pisane* is limited to the Mediterranean and the Black Sea and down the Atlantic coast, and it is a map of directions for sailors. The most famous quasi-portolan, and the one that might have had some influence on Fra Mauro, is the *Catalan Atlas*, said to be drawn in 1375 by Abraham Cresques, a Jewish cartographer working in the Catalan tradition on the Spanish island of Majorca. There was a famous group of cartographers working there and most of them were Jewish, as were many of the sailors who traveled to Africa to pick up trade goods.[90] Cresques used travelers' reports, especially the stories of the Venetian trader Marco Polo, to draw his map.[91] It is made up of six double velum panels, each about two feet square in size, and it folds up, like a screen. The first two double panels are covered in text and illustrations; the text is a medieval geography lesson. The second panel is a magnificent calendar wheel with information about the phases of the moon, the place of planets, signs of the zodiac, and a way to calculate tides. The map of the world is both a mappamundi and a sea chart. The map stretches across four of the panels and is a map of the known world, including the various monsters, myths, and strange creatures that appear on other medieval maps of the time. There are no blatantly religious overtones, nor paradise or the Garden of Eden on this map. It seems that some Judaic thought was incorporated into the *Catalan Atlas*, which makes sense if Cresques was the cartographer. It is also overlaid with the typical portolan rhumb lines used by sailors. These lines clearly mark trade routes to Africa, but this map would never have been taken aboard any ship for navigational purposes—it was a present for the king of France—because it was a decorative piece that combined mappamundi traditions and portolan charts. [92]

One of the great portolan cartographers was the Italian Pietro Vesconte, who was born in Genoa but worked in Venice as an adult in the early fourteenth century. We know of Vesconte's work because he was the first cartographer to sign and date his nautical charts. [93] On one, dated 1318, there is a small portrait of a gentleman in the upper right corner sitting at his drawing table, and under that picture is Vesconte's signature. Presumably, this image is a self-portrait of the cartographer at work. Vesconte also produced several atlases of portolan charts. In 1321 he made an atlas of nautical charts and a world map for diarist Marino Sanudo, who was trying to convince Venice to launch an eleventh Crusade. [94] This eastern-oriented world map is a circle, drawn on two pages, and it is also overlaid with the typical rhumb lines of a portolan chart. Anchored at sixteen points around the edge of the map, these lines then stretch over all the surface, across land and sea, which distinguishes this map from other mappamundi because it makes this map look more like a chart. These lines also have absolutely no relationship to the grid of latitude and longitude, and some of them also meet in the center, which is Jerusalem. [95] The continents on the Vesconte world map, in white, take up much of the world. The landmasses are encircled by green oceans, mountain ridges are marked in bumpy brown, and there are some red symbols. Most interesting, many of the few inscriptions on the map are placed along coastlines, which is typical for portolan charts.

ITALIAN WORLD MAPS AND THEIR POS-
SIBLE INFLUENCE ON FRA MAURO

It's a reasonable assumption that Fra Mauro would know about maps made in his region. Of course, there was no unified Italy back then, but close geographic proximity, economic entwinement, a common religion, and a common language—Latin—united northern Italy culturally, even in the face of political divisions. There was also continuous communication among citizens, merchants, diplomats from city-states, and the Republic of Venice as people traveled about.

An Italian atlas was made around 1351 (or perhaps 1371) called the *Medici Laurentian Atlas* after the grand Florentine family. It was probably drawn by a cartographer in Genoa and consists of eight pages, the first an astronomical calendar and the second a world map. The rest are portolan charts. What forms the world map so special to the history of mappamundi is that it has an identifiable Africa shaped much like the real Africa, although it is top-heavy where it defines the Gulf of Guinea correctly, it forms the rest of Africa as disproportionally skinny. More significantly, this world map shows the tip of Africa surrounded by water and that makes it the first Western representation of the possibility of circumnavigating Africa. But no one knows who drew it, why they chose to draw South Africa surrounded by water, or what effect this picture might have had on others, especially traders, sailors, and explorers. It's also notable that Africa has no place names past Cape Bojaor (in present-day Western Sahara, which is controlled by Morocco and is across from the Azores), a point where most navigators turned around and went home. The absence of toponyms in the southern part of the continent might be a sign that this part of Africa was imagined, and that our surprise at the water surrounding the base of Africa should be tempered.

The world map made by the Venetian Albertinus de Virga around 1413 has some superimposed lines, but unlike a portolan, they simply divided the world into eight equal pie slices. This world map is also signed. Although it was lost in the twentieth century, there are many photographs of the original. Virga's circular world is only about 16 inches (40 cm) in diameter and set on a long, decorated parchment that still has the flattened neck of the animal that once wore that skin. It has the usual three continents. This map's Africa is also free-floating, not attached to the edge or to any other landmass, which makes the case for the possibility of circumnavigating Africa. Outside the map are two charts, one for calculating the phases of the moon, the other for figuring out the date of Easter over the years, a date that was important back then. Unlike so many other mappamundi, this one is oriented north. It's also a very colorful map with yellow lands, white seas, brown rivers and mountains, and blue lakes. The Red Sea stands out because, unlike other waters, it is red. This map also has a large and artistic circle of the signs of the zodiac on the neck of the parchment.

About twenty-five years later in Venice, in 1439, Andrea Bianco, who was an experienced sailor and cartographer, drew another Venetian map of the world. This map is extremely important to the story of Fra Mauro's map because Bianco collaborated on that map (see chapter 4). Like Vesconte's world map, this one was part of a nautical atlas of ten pages called *Atlante Nautico*, which he signed and dated. He also includes instructions and a table full of trigonometry on the first page if others wanted to find their course using his map and a compass.[96] Bianco's Earth, included as Table 9, is about 9.5 inches (24 cm) in diameter with a thick circle of blue surrounding the green oceans that circle the land. It runs across two pages, filling up the left page more than the one on the right, and is a mappamundi in its construction and style. Bianco's world map is oriented east, which might have been an interesting discussion for Fra Mauro when he was deciding how to orient his map fifteen or so years later. The southern tip of Africa extends into the blue that frames the world and this too is unlike Fra Mauro's ideas about what to do with Africa. Bianco's map also has eight radial lines evenly spaced across the circle of the earth, as in de Virga's, and they also divide the world into eight equal sections. The land is inscribed with text and figures, and it has mountains and cities. The Red Sea is misplaced but colored red. This mappamundi also has the traditional icons of Christianity, such as paradise and the four rivers, the Virgin Mary and Baby Jesus, and Noah's ark. There are the usual human monsters, such as mermen, dog-faced men, and wild creatures, including elephants and dragons. None of that shows up on Bianco's nautical charts. This was Table 9 of his atlas, and the final Table 10 is another world map modeled after Ptolemy's projection. That map is spread out in the cape shape typical of Ptolemaic maps; it has curving latitude lines and evenly spaced longitudinal lines that fan from the top and then drop together over the equator. Historian Andrea Ferrar has suggested that Bianco probably had help with this map because the handwriting is not his, and it might have been drawn later after the rest of the atlas was completed. It could also be a copy of some other Ptolemaic map.[97] Taken altogether, this atlas sums up medieval cartography and sets the stage for Fra Mauro's map.[98] We also see here an experienced and accomplished cartographer, one who had already contemplated and produced a mixture of three kinds of maps. Bianco was just the person to guide Fra Mauro.

Another cartographer, Giovanni Leardo, made three world maps in the mid-fifteenth century in Venice. Three copies, dated and signed by Leardo, have survived.[99] There are rings around the world on all three maps, and one has three rings. Within the rings, all three have the same basic inner world form—a circle of Earth oriented east with Jerusalem in the center, but they differ in their simplicity or decoration, and the number and function of the outside rings.[100] They all have landmasses in red and brown to the north and south, representing unknown territory. Africa is attached to the southern empty land, and Europe is connected to the area in the north with a warning band in between stating that this part of the world cannot be inhabited because it's too cold. The water on all three is a wash of blue, and the land is white and covered with red buildings and clusters of buildings that represent cities. Leardo also put paradise at the top of his maps, Noah's ark on Mount Ararat, and the scary fictional monsters Gog and Magog in the top corner. His text explains all sorts of myths and legends usual for medieval times. For example, Leardo warns of flesh-eating people and scary griffins. His most accurate cartographic work is, of course, the area around the Mediterranean, which Italian sailors knew well.

Another Italian world map from about the same time comes from Genoa, on the west coast of Italy. Dated 1447, it has no author. The most fun part of this map is that it was drawn lozenge-shaped, a long oval that has points on either end. The map is spread over two pages, and it is oriented north, like maps drawn today. We know it was made in Genoa because there is a Genoese flag planted on the parchment and a coat of arms of a Genoese noble. It also seems to follow the tales of Venetian explorer and traveling merchant-trader Niccolò de Conte. [101] And it is covered with plenty of monsters. Rumor has it this map was used by Christopher Columbus as he sailed west.

HERE BEGINS FRA MAURO

Historian of cartography David Woodward suggests that there was no such thing as a cartographer in the Middle Ages. It was neither a job nor a pursuit, not a discipline at all. Rather, artists, tradesmen, monks, and even common citizens drew maps, often for practical purposes. And many maps were inserted into texts and served just as illustrations. [102] But taken together, these mappamundi, and the portolan charts, sum up the knowledge of world geography before Fra Mauro picked up his inks. Word maps, from every culture, became more accurate and geographically inclusive over time for obvious reasons. As sailors and traders moved about the seas and oceans, they collected more precise knowledge of "foreign lands," information which then filtered back to cartographers and their patrons as well as the merchants who would be commissioning more practical portolans. And so, cartography improved as navigational skills improved and shipbuilding technology provided faster, bigger, and better vessels. [103] In the process, mappamundi were released from the typical T-O form and moved into the realm of geographic reality. [104] At the same time, a better understanding of astronomy and the heavens, upon which much early cartography relied, also improved because of inventions such as the telescope, which Galileo Galilei perfected while working at the University of Padua near Venice from 1592 to 1610. [105] In Western culture, however, world maps were also more dogmatic than instructional and designed as art objects that promoted Christian dogma, and less practical than regional maps or portolans, which were used for navigation on land or water.[106] Instead, mappamundi were designed to give the viewer a sense of the geographic world, his or her place in that world, and in the West that place that was geographically secured by religious adherence to a belief system centered on the Bible and Christ. These maps were instructive and informative, but most of all, they were reassuring, because they presented a world organized by an accepted theology. The purpose of religion is to provide a worldview that bonds anxiety and makes for a collative identity based on faith in a particular belief system. Religion also dictates a way of life, and that structure often calms the busy and anxious mind. And so, it's not surprising that the mappamundi of medieval times, especially in Christian Europe, were revered documents viewed by the faithful who had no experience with worlds beyond their own.

But lurking in some of these maps was the possibility of another world, where God was not in charge and all those monsters and weird creatures at the edges of the world might turn out to be interesting cultures yet to be known and understood. After all, the Middle Ages was transforming into the Renaissance in the West, and the maps leading up to that period hinted at a coming transition in the style and purpose of mappamundi that presaged the Renaissance. The Age of Exploration or Discovery had begun, and so it's no surprise that expansive portolan charts were necessary as practical guides for exploration, trade, and expansive colonization. For world maps, it was a time for listening to mariners and traders and then making mappamundi that stood on their own rather than serving as illustrations for manuscripts and atlases.[107]

And then came Fra Mauro's map, the instigator of change, the map that rejected religion and went so far as to embrace the nascent methodology and philosophy of science, thereby changing both maps of the world and the world itself. Although Fra Mauro's map follows much of the traditional mappamundi format, it has tweaks and new approaches that also make it not just different from the long list of typical mappamundi that went before. His map was on the edge of blasphemy.

CHAPTER 3

The World of Fra Mauro

If you have just come to the monastery, and in spite of
your good will you cannot accomplish what you want,
take every opportunity you can to sing the Psalms in your
heart and to understand them with your mind. And if
your mind wanders as you read, do not give up; hurry
back and apply your mind to the words once more.
　　　　—Saint Romuald, founder of the Camaldolese order,
　　　　　　　　　　　　　　　　　　　　(957–1027 CE)

Many of the graves here seem disturbed,
As if the owners had vacated suddenly,
Hearing the high-tide warning ringing out'
Sick of the damp, poor plumbing, spreading mold . . .
To get to the outskirts, where the poets lie,
you have to pass the teeming suburbs of the dead:
broad avenues of concrete tenements,
anonymous but well-maintained,
each with its little window box of artificial flowers. . . .
　　　　—J. Ziguras, *Isola di San Michele*

Today, Venice is a bustling city. Combine the fifty thousand or so residents with the thirty thousand tourists that stream in every day, with everyone walking about among the medieval labyrinth of tiny streets because there are no cars, and what you get is something like a massive, swirling, loud, outdoor festival of humanity on the move. But there is another side of Venice that few tourists, but all the residents, cherish and know well. Venice has many quiet parts far away from the historical center. Go east to the *siestere*,[1] or district, of Castello and find Venetians going about their everyday lives. Walk to the west and mingle among students of Ca'Foscari (the University of Venice) or the architecture and design school Università IUVA di Venezia, eating takeaway pizza and drinking beer in Campo Santa Margarita, making that part of the city much like a college town.

Or go north, to stand on the very edge of the city on the wide Fondamente Nuove, where the public waterbuses come and go, dropping off people as they circle the city or branch out to the various islands of the lagoon. If you do this early in the morning, or at dusk, and pause to look north over the Venetian Lagoon spread out in all its glory before you, the city takes on new meaning. Yes, there will be boat traffic as fisherman and cargo transport boats go about their business no matter the time of day, and the water buses will approach and leave on schedule, and yet, this view is one of the most lonely, meaningful, and melancholy that Venice has to offer. The gray waters of the shallow lagoon, stretching seemingly forever, are accented with rows of wooden pilings (*brocole* in Veneziano) that since 1439 have marked out channels where boats can travel and escape the shallower parts of the lagoon.[2] Meant to keep boat traffic in their lanes, as roads do on land, these limbed oak stakes are banded together with metal strips or links of chain and often topped with resting seagulls and other birds, or decorated with shrines to loved ones lost at sea.

The view is expansive, open, and inviting for those who might be curious about what else this region has to offer besides the tourist-packed historical center of Venice. And close by, centered in the view, but accessible only by boat, is an island called San Michele in Isola but better known as "the island of the dead." Five and a half centuries ago on this small offshore island, Fra Mauro drew his map.

SAN MICHELE IN ISOLA, THE HOME AND ISLAND WORKSHOP OF FRA MAURO

Calling San Michele[3] "the island of the dead" is not so much a nickname as a fact. This is where Venetians today are buried, but sometimes not for

long. Because there is so little ground in the city, Venetians have always
had to figure out what to do with cadavers. Up until the nineteenth century, they simply dug up some paving stones in the city and placed the
bodies underneath; these streets were called *campielle dei morti*, or streets
of the dead. But still, with a city population of about 180,000 people
at its height, there were not enough paving stones for those bodies to
remain forever buried in the city. Taking advantage of sitting in a lagoon
full of islands, the city then acquired abandoned islets and turned them
into ossuary islands—islands for bones. Bodies interned in the city were
then routinely dug up after a few years and the bones rowed en masse,
and without individual identification, to these islands and dumped in
a pile.[4] Even today, these islands are empty of live people, secretive,
only accessible by boat, and known only to Venetians. For example, the
long-abandoned island of Sant'Ariano was appropriated by the Venetian
government in 1565 and was used continuously as the last resting place
for many Venetian citizens for three hundred years. It was shut down
in 1933, and by then the mountain of bones was three meters deep and
overgrown with vegetation.[5]

The island of San Michele is now the place where Venetians are
taken after death and some remain there forever while others are, as
per custom, dug up after twelve years and taken elsewhere. Standing
on Fondamente Nuove and looking straight across the water at San
Michele, this island has a distinctly ordered look that contrasts with
the hodgepodge of other Venetian architecture seen in the city proper.
Instead of a mix and match of Gothic, Byzantine, and Baroque styles,
the island is designed in a more classical mode which suits its solemn
purpose. The entire island is enclosed by a simple high brick wall
trimmed in white, and etchings from centuries past show that there
has long been some sort of wall enclosing this island. Now, there is a
central gate on the south side facing Venice, but this entrance is only
used during a special festival when the city sets up a temporary bridge
so that mourners can walk from Fondamente Nuove to the cemetery.
That floating bridge used to be constructed each year on November 1,

All Saints Day (Festa di Ognissanti), and the bridge was called Ponte Votivo for the occasion. The practice was suspended in 1950 but was reenacted for the first time in seventy years in 2019.[6]

The formal gate to San Michele on this south wall boasts two rounded tower-like ends with white spikes sprouting from the rooves. There is a three-arched door entrance in the middle and white stone steps lead down to the water. This gate looks more like the entrance to a castle than one built for funerals, but the winged angel, the Archangel Michael (San Michele) that stands at the peak of the central triangular roof as well as the two crosses atop each tower mark this as a spiritual, holy place. Tall slim cypress trees rise above the wall inside the cemetery and act as verdant sentinels of the sacred duty of this island. From the shore of Venice, all you see of San Michele is the wall, the gate, and the row of very tall cypresses.

The pictorial solemnity of the island is further accented by a metal sculpture by the Russian artist Georgy Frangulyan. It sits just offshore from the gate and is titled *Dante's Barque*. It consists of a small boat that bobs in the water in response to passing waves. Standing in the boat, as Venetians do, are medieval figures representing Dante and Virgil, who are approaching San Michele, presumably as the entryway way to the underworld. One of them, apparently Dante, has his arm raised, pointing to San Michele as if pointing the way to hell or heaven that lies through San Michele. Erected in 2007, it feels as if this sculpture has always been there, ever since San Michele transitioned from a monastery island to a cemetery. The work is beyond prophetic; passing by on a waterbus it's impossible not to feel a shiver right before the boat pulls up to the landing stage Cimitero on the island's western side.

Once on the island, a visitor walks though the entrance into the cemetery, which takes up most of the island. Depending on your sensibilities, the view is either creepy or comforting. Famous outsiders such as Igor Stravinsky, Serge Diaghilev, Ezra Pound, and Josef Brodsky are buried here and they have not been moved, but the population is overwhelmingly

Venetian. Without any of the usually jam-packed Venetian buildings crowding the way, this island retains its verdant greenery swept by lagoonal breezes. It looks more like a park than a cemetery. Further along are newer sections of high walls where bodies or cremated remains are put in drawers and there is a sliding ladder, as in a library, to reach and maintain, or decorate, those individual graves. On all the newer ones, there are real or plastic flowers in vases, electric votive lights, mementos, and photographs. In 2018 an article in the *New York Times* claimed there were eighty-five thousand people buried here, which makes this deceased group now larger than the population of Venice, which is about fifty thousand people.[7]

But turn a sharp left after the entrance gate and enter the part of San Michele that is still a church run by monks. Immediately, you enter the original cloister of the monastery from the thirteenth century, with its rugged stone floor. Gravestones on the walls and under your feet echo names of long-past illustrious Venetian families—Vianello, Zorzi, Contarini, Grimani. There are thirty-nine family chapels in the arches in this part of the cloister, and they are usually closed off by bars, but you can look in.[8] Continue on and enter a second cloister called the Small Cloister (Il Chiostro Piccolo), built in 1436, which is a sun-filled loveliness of arches and columns with a large central wellhead set on three steps and crowned with a curly iron bow.[9] This cloister has a second story above the tiled roof that runs on all sides with windows spaced at intervals. Presumably, this is where Fra Mauro lived and drew his map along with other monks of the Camaldolese order.

THE HISTORY OF SAN MICHELE AND THE CREATION OF THE CAMALDOLESE ORDER

The island of San Michele was not always the cemetery island: it has gone through various permutations before gaining its current look and function. The island was initially just an isolated bit of land set between

Venice and the island of Murano, presumably covered in the typical wild vegetation that today overgrows abandoned islands in the lagoon. Then, according to an unverified story, a wandering holy man named Romuald or Romualdo (951–1027 CE) from the city of Ravenna spent ten years living there with a hermit and holy man named Marinus, who had been the only occupant of the island before Romuald arrived. There were all sorts of hermits in Italy at that time and Marinus was well-known as a "spiritual master," a person who held the tenants of a hermetical life close, but we know little about his background. Romuald had come to the path of a hermit after witnessing a duel between his aristocratic father and another man. Even though that duel had nothing to do with the young man, Romuald took it on as his own sin. That act transformed him from a noble rake who had been engaging in all the pleasures of life into an aesthetic, religious holy man. After the duel, the young man entered a Benedictine monastery in Ravenna to reflect on and repent his part in the duel, but that monastery was less than conducive to a life of reflection; compared to the other monks, Romuald was more dedicated and pious and he didn't fit in.[10] And so, three years later he fled to Venice, attracted by the reputation of Marinus. Life on an island with one other hermit would allow Romuald peace and space for the reflective life he desired. Years later, and far away from Venice, Romuald founded the order of Camaldolese monks, and he is now known as Saint Romuald.

Not much happened on San Michele for two hundred years after the two hermits lived there, but upon Romuald's death in 1027, it made sense to honor his early life and the decade of living with the hermit on a lagoonal island of Venice. Romuald's monastic order, the Camaldolese, eventually gained the rights to the island. They then established a monastery there in 1212. They also built a church dedicated to Saint Michael the Archangel, which is why the island henceforth carried the name San Michele. That original church was consecrated nine years later, and then renovated in 1392. Only the original cloister from the thirteenth century that was part of the original monastery or hermitage remains intact. The

monastery also held a precious cross that had been stolen from Constantinople by conquering Venetians, but this relic is now at a hermitage in
Fonte Avellana in the province of Marche.[11]

The community of Camaldolese monks on San Michele was famous.
It eventually become a grand center for learning and its library was
renowned among intellectuals. San Michele was so influential in its
time that it annexed, or connected with, four other monasteries around
the lagoon, making San Michele the center of cenobitic, or monastic,
life in the region. It was also the first Camaldolese monastery that was
allowed to elect its abbot without approval from the prior general, also a
signal of its competence, authority, and singularity.[12] Historian Angelo
Cattaneo says that in the fifteenth century, San Michele was "an ecu-
menical crossroads of men, knowledge, and cultures."[13] The monks had
a reputation for studies in philosophy, science, and the arts, and there
was a school that taught the art of miniature painting. They also trained
artists and scribes to illustrate and copy manuscripts. By 1612, the monastery of San Michele broke somewhat from the Camaldolese order and
became their own cenobite community, which seems to suggest there
was less time for, or interest in, the eremitic part of Camaldolese life.
Or maybe that decision was anchored in the usual Venetian sense of
independence that cried out for local rule. This separation was resolved
in 1935 when Pope Pius XI reunited the San Michele Camaldolese back
with the rest of the order.[14]

The island remained as it was, with its original 1221 church, monastery, and cloister, on the west side of the island for almost a hundred years.
There was a renovation in 1392 and then a fire in 1453 that destroyed the
church. The architect Mauro Codussi was hired to design a new church
for the island to replace the previous house of worship, and he did so to
honor the Camaldolese order and their beliefs. The new construction
was finished in 1469. The design was revolutionary because of the three
simple front panels made from bright white Istrian marble and incised as
if made from bricks. With its central front door that opened to the water,
arched skinny windows on each side of the door, and large oculus window

above, this small church is considered the first Renaissance building in Venice and the introduction of classical architecture to the city.[15] Built right on the water, this gem of a church sparkles with the light coming off the waves of the lagoon. Historian of Venetian architecture Debora Howard suggests that the all-white façade, as opposed to most buildings and churches in Venice, which were decorated with varied colored marble, makes the church of San Michele seem like it is "floating like an iceberg in the lagoon."[16] Inside, the church is filled with polychrome marble and high gray and white arches with windows at their apex that let in light. It is also a working church with areas for meetings, offices, and priests' garments hung on a rack. Attached to the church, and also on the waterfront, is a round and tall chapel called the Capella Emiliana, designed by Guglielmo Bergamasco and built in 1530 as a six-sided small chapel full of polychrome marble. Behind the church is a small bell tower made of brick from 1460; significantly, it was not pulled down when the Codussi church and the chapel were built.

Three hundred years later, Napoleon Bonaparte arrived in Venice and suppressed, meaning closed, the Camaldolese monastery on San Michele in 1810, a time when he also closed so many monasteries and convents in Venice. Acting on Napolean's orders, the interiors of these places of God were stripped, their sacred objects and prize statuary sold, and the money sent to France. Luckily, items from the renowned San Michele monastery library were transferred to Venice and many of them ended up in the Archivio di Stato (State Archives, meaning the archives of the Republic of Venice, not the Italian State) and the Biblioteca Nazionale Marciana, the city library. That transfer is why Fra Mauro's map, which hung in the church on the left side of the nave from 1459 to at least 1810, ended up at the Biblioteca Nazionale Marciana which continues control where it hangs today.

Napoleon is also why San Michele became the cemetery island for Venice. When the French occupied Venice under Napoleon's army starting in 1807, burial within the city was banned by the occupiers. They decided that this practice was unsanitary, especially since the waters of the lagoon sometimes swelled the embankments and seeped under the paving stones. But some archaeologists who have worked on unearthing evidence of graves in Venice have suggested it was more a prejudice against Venetians and their unusual death customs that offended the French and motivated the establishment of a cemetery island.[17] In 1807 the French Imperial Army ordered that bodies must be taken instead to a small island off the north side of the city, an island that was very close to San Michele, called San Christoforo (Saint Christopher), and be put into the ground. That project was directed by the official city architect Giannantonio Selva, who demolished the church and monastery of San Cristoforo in the process, which, until then, had been one of the most notable landmarks of the northern view off Venice.[18] The first funeral was held on San Cristoforo in 1813. Twenty years later, there was another pressing need for graves, even though bones were being routinely dug up and moved to faraway ossuary islands. And so, the occupying forces filled in the narrow canal between San Cristoforo and San Michele, bundling them together under the name San Michele as the cemetery island of Venice.[19] They then disinterred people already buried in Venice and moved them to the newly constructed joint cemetery.[20] The first map of San Michele as we see it today dates from 1846. But in the process of digging up the city graveyards, which were attached to the various churches across town, the French also dramatically changed the cityscape of Venice. Where there had been a spatial connection between the living and the dead buried under the pavements in every small neighborhood, now there were empty spaces left by the disinterred graves. The empty and spacious *campos* that surround these churches today seem light and airy, but these spaces are not being used as they were intended, as gravesites.[21]

In 1872 the cemetery island received another makeover, designed by engineer Annibale Forcelli. At that stage, the island was enclosed by the high brick wall, planted with cypresses, and divided into various areas or plots, including special places for those of Greek Orthodox and Protestant faiths. In 1998 the municipality felt the cemetery was running out of room again, even for short-term burials. British architect David Chipperfield then won the contract for an extension of the northeast corner of San Michele, where landfill helped increase the available space for bodies.[22]

A few Franciscan friars took over the island at some point, and they lived there until recently. Now, the priests are only day-trippers like everyone else, except the dead. Their job is to oversee the cemetery, help with funerals, and comfort the mourners.

WHAT IT MEANS TO BE A CAMALDOLESE

Fra Mauro would likely have identified himself as a Camaldolese monk who lived at the island monastery on San Michele. What does it mean to be a Camaldolese and what did it mean in the fifteenth century to have that identity? The answer to that question goes back to the time St. Romuald spent on that island in the Venetian Lagoon with the hermit, long before there was a monastery or a cemetery.

After ten years living on San Michele with the holy man Marinus and learning from him, Romuald moved with Marinus and another newly minted monk who used to be the doge of Venice, Pietro Oresolo I, to Cuxa in Catalonia, which we now consider part of Spain. Romuald and Marinus soon built a separate hermitage there near an already established monastery, always seeking the most extreme version of solitude and aestheticism.[23] Eventually, the monastery and hermitage were connected and Romuald was anointed head of both. After ten years in Cuxa, Romuald set off across Italy, and for thirty years, he evangelized his style of spiritual devotion and reform while mostly living the solitarily life or staying in monasteries. Romuald's goal was to reform other monasteries into a stricter vision of the hermetical life. He also wanted everyone else, the general population that is, to be like monks and take on a contemplative life.[24] Romuald spent time in Rome, then he went back to Ravenna, and eventually, set off wandering again, landing in Arezzo, in Tuscany, in an area called Camaldoli, around 1015. He then acquired some land and built a small hermitage, which became of founding place for a new order of monks established by Romuald. It was called the Camaldolese order in the spirit of the area, Camaldoli.[25] The need for a new order was underscored by a vision that Romuald had of monks dressed in white climbing a ladder to heaven; the Camaldolese thereafter have dressed in white.[26]

The Camaldolese order was based on the Benedictine model and the Benedictine Rule, but it was meant to be stricter. While the order was, and still is, closely aligned with the Benedictines, it has been recognized as separate since Romuald's death in 1027. This order has always been small, but some of its members have had great influence. For example, the monk Guido of Arezzo invented the *do-re-mi* scale we use today; the Jurist Gratian wrote *Decretum* in the twelfth century, which is the basis for canon law; and in the fifteenth century Nichlas Malerbi published the first translation of the Bible into Italian. The order also includes the renowned humanist Ambrogio Traversari.

The Camaldolese life is a quiet one, based on solitary contemplation and meditation, but that hermetic life also includes some time working with others in a monastery. In other words, Camaldolese live a combination of the cenobite (communal) life and eremitic (solitary) devotion. Followers ascribe to the need for solitary time, fasting, and silence, but also realize the benefits of being part of a religious community that lives and works together, the Benedictine part. That philosophy was the driving force when Romuald built his first hermitage near a Benedictine monastery in Tuscany. Yet Romuald was not so keen on the monastery part of this life and more in favor of the solitude of a hermit. He also didn't necessarily follow the rules of the monastery abbot but thought of his hermitage as complementary to whatever the Benedictines were doing.[27] Since the housing on San Michele has been always referred to as a monastery and not a hermitage, we can assume that while Fra Mauro surely had time for singular reflection, he was not a hermit per se.

The Camaldolese essentially follow the Rule of Saint Benedict from 516, when the saint wrote that members should live in a community with an abbot as the leader and practice a life that combines work and prayer.[28] The Benedictines have been successful because they did not embrace religious zeal and they recognized that it is possible to have a balanced life full of devotion to God and yet also interact with others in an orderly, peaceful way in a confined community.[29] Romuald added to that basic structure the idea that there should also be solitude away from others and from the culture at large. It was only through such eremitic practice, he felt, could one speak with God. So Romuald was a hermit and yet he roamed the world and interacted with people while also setting up monasteries and hermitages. In the same way, Fra Mauro lived in a monastery within which he would have had plenty of time to be alone, but he also has the opportunity to socialize with his fellow Camaldolese and with the outside world as well in his role as a cartographer and other jobs. From the Camaldolese view, time spent with others in the monastery might be considered a second stage, where a hermit was devoted to prayer but might also spend time with like-minded faithful. Interacting with those outside the monastery provided another social layer and a chance to share the love of God with others as missionaries. Overall, Camaldolese believe that life should be entwined actively with the word of God, gathered from the Bible. They also believe this pursuit should be an all-encompassing reflection and expression of God, a sort of "living witness" to the glory of God.[30] This kind of life is "living together alone," in a combination of "solitude and communion," all under the eyes of God.[31]

Camaldolese and Benedictines also put great store in love. As one theologian put it, love is "the durable seam" that ties together space and time.[32] In their minds, God is love, and that love should be spread among as many people as possible. For monks, that love begins with their brothers in the community and their obedience to the abbot; everything is done in the positive light of love. Such love involves compassion, respect, understanding, and friendship. For Camaldolese, it also means love is extended to other religions and they are known, even today, for making bridges across faiths.[33] Countries and cultures based on the economics behind capitalism, which encourages the individual to look out for her or himself, procuring individual wealth at the expense of others, is anathema to the Camaldolese and Benedictine ways. Given that approach, it must have been difficult for a monk such as Fra Mauro to spend the day in the solitude of love and community and then walk about in a city such as Venice, an engine of capitalism and individual monetary gain.

WHO WAS FRA MAURO?

We know little about who Fra Mauro was.[34] We don't even know his real name. Mauro is a common Italian male name, and it has a bit of connection with Benedictines, but for all we know, his first name really was Mauro.[35] It's just as plausible that he changed his name to Mauro when he took the vow of a lay (or secular) monk. There is also no record of his original last name or his birthplace beyond the fact that it was somewhere in Venice. Some have suggested he was from the nearby island of Murano because San Michele is often referred to as "San Michele in Murano." San Michele is the islet that sits between Venice and Murano, and "in Murano" translates from Italian to English as "part of Murano." In essence, that "in Murano" is just a sort of postal address. These two islands are separated by a three-minute boat ride, but they have had separate functions forever. One is home to Venetian glassmaking while the other is where Venetian are buried. One is populated today by five thousand or so Muranese while no one lives full-time on San Michele. Also, there is no church or state record that Fra Mauro was born on Murano or anywhere, but that's probably because no one knows his real name.

It's just as frustrating that there is no evidence of who Fra Mauro was before joining the monastery on San Michele, why he chose that order, or why he remained a secular monk (*converso*). There were certainly many other monasteries in Venice representing various orders and he could have had his pick. Records do show this sentence, dated July 9, 1409—*Frater Maurus de Veneciis conversus*—which acknowledges the presence of Fra Mauro of Venice who was a *converso*, but who knows if this line even applies to the monk who made the map.[36] There might have been another Fra Mauro at the monastery. But there is an additional entry in the registry of a monastery, dated April 15, 1434, which says that Fra Mauro of Venice was a member of the Chapter of San Michele, which makes it definitive that he was there.[37] But we do know someone named Fra Mauro made the map since this name is later found in monastery financial records noting payment for supplies and helpers for making the map (see chapter 4).[38]

The mystery of who Fran Mauro was is confounded by his remaining a lay monk. As historian and expert on Fra Mauro's map Pietro Falchetta says, the addition of the word *conversus* is ambiguous at best. It might be that a *converso* was less than a full monk, or that he was not of noble birth, or maybe the term referred to monks who joined as adults rather than as boys, a time when young males were often put into monasteries by their families. Embedded in that confusion about his position as a *converso*, also known as "lay brother" or "secular monk," is the question of whether Fra Mauro was a fully ordained priest. He was without a beard on his posthumous commemorative medal and only ordained priests were clean-shaven. One motivation for not being ordained and remaining at the status of converso meant Fra Mauro could easily interact with the outside world while the other friars were more often restricted to the monastery. Documents show that Fra Mauro was given jobs that historian Angelo Catteneo calls "humble but necessary," and these tasks required leaving the monastery and having conversations with outsiders, a privilege that would serve him well as a cartographer.[39] Records show that Fra Mauro had the job of collecting rents on the various properties that the monastery owned, presumably in Venice. Also, in 1437 he was tasked with mapping the Camaldolese monastery and surrounding area in Istria (Croatia) at San Michele di Leme, on the other side of the Adriatic.[40] The specific problem Fra Mauro was trying to solve with this map was a property dispute between the monastery and its neighbor. That map was further engraved and printed in the eighteenth century because it was disintegrating.[41] It now only exists as a reproduction in the Camaldolese journal, but even that reproduction signals that Fra Mauro had cartographic skills, that he was interested in mapmaking, that he could do drawings of topography and borders, and that his cartographic skills were recognized and trusted by higher-ups in his order.

Six years after the Istrian assignment, in 1444, Fra Mauro is listed as having another job.[42] He was appointed by the Savi ed Esecutori alle Acque, the government entity that took care of the lagoon and its many confusing water issues. Of course, this ministry would be essential for Venice, a city that had been rerouting rivers and worrying about silt buildup in the lagoon ever since its founding. In this case, they were seeking Fra Mauro's knowledge to change the course of the Brenta River, a major river that runs from Trentino in the Dolomite Mountains down into the lagoon just south of Venice. Although there is no record of Fra Mauro as a hydraulic engineer, he was invited to help map the area and the course of the Brenta in the role of cartographer. That reputation was also possibly why he was called upon to witness a document in 1445 that put forth an agreement between the Venetian government and the patriarch of the city of Aquileia on the mainland that involved a boundary dispute.[43] These few documents at least show that Fra Mauro had a reputation within his order, and with the government of Venice, as someone who looked at and drew land and man-made features to scale using the accepted symbols of topography. In other words, they were certain he could understand what geography meant in terms of ownership and boundaries. Fra Mauro's work with the water committee also points to his understanding that rivers, land, and everything on Earth, were subject to human intervention and change.

Historians presume that Fra Mauro died sometime in October 1459 before the map was framed and hung; the map is dated on the back of the frame as August 26, 1460. They offer a death date of the fall of 1459 because, on October 20, 1459, there is a record of a small wooden chest containing Fra Mauro's writings and preliminary drawings of the map being transferred from San Michele to the monk Don Andrea, the abbot of the monastery San Giovanni Battista. That Benedictine monastery used to be on the island of Giudecca which lies just south of Venice. Maybe Fra Mauro was ill, and he held on to his papers when he was transferred to the other monastery for care and then he died, and the papers remained. We do know that five years later, that box was returned to San Michele. That story appears in a book about Fra Mauro's map by Placido Zurla, also a Camaldolese monk, which was published in 1806.[44] By then, the writings and drawings were long gone. Historian of Fra Mauro's map Angelo Cattaneo claims that the loss might not have been a case of something being inadventently mislaid. Instead, they might have been purposefully thrown away. It seems that by 1480 the Camaldolese had removed all of Fra Mauro's writing from the monastery and obliterated his name from the work of the map. Also, the general archives of San Michele were thereafter moved repeatedly, which meant many of the documents that might pinpoint what happened to Fra Mauro's box, or anything with his name on it, were probably lost in all that moving about.

Even with this information about the monastery and the Camaldolese order as his belief system, lifestyle, and orientation, it is still difficult to imagine what Fra Mauro looked like because there are no drawings or paintings of his image. Perhaps appropriate to his role as cartographer, scholar, and devout monk, such recognition would not have been important to him or anyone else. No matter that we praise his world map in hindsight, Fra Mauro was, in essence, a cartographer for hire; he certainly did not personally sign his great map as an artist would have. Or maybe his focus was on the map and not the maker. There is only one image of Fra Mauro, and that one is on a commemorative coin attributed to artist Giorno Giovanna and struck after his death. Fra Mauro's profile on that coin shows an old man wearing a pointed skull cap that completely hides his hair and ears and a monastic hood draped around his shoulders. Since the coin is not full-faced, we have no idea of his expression, or what he really looked like.

In other words, the creator of one of the greatest maps in the world remains an enigma.

Figure 1: Fra Mauro's Mappamundi, 1460, Venice, Italy.

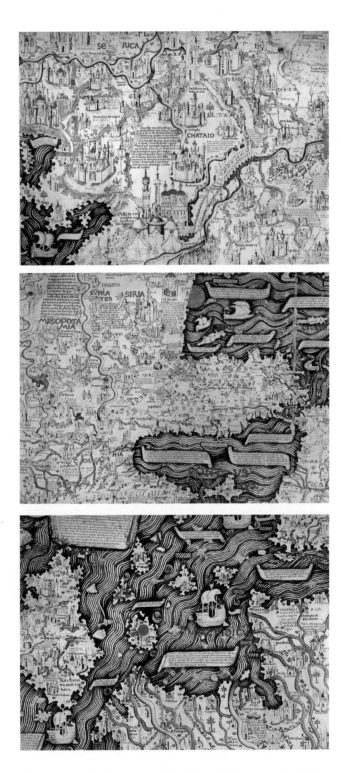

Figures 2–4: Enlargements of Fra Mauro's Mappamundi, 1460.

Figure 5: The cemetery island of San Michele off the northern shore of Venice where Fra Mauro lived and drew his map. Long after his death, two islands were joined to make this larger island and it was then designated as Venice's cemetery island. The surrounding brick wall and the cypresses were added. In Fra Mauro's day there was a monastery complex and a church. The church has been rebuilt but the cloister remains.

Figures 6–8: Since 1823 when San Michele was joined with another island nearby, it has been the Cimitero di Venezia (cemetery of Venice), including burial and cremation. Since space is limited today, many burials are temporary and remains are often removed to other islands after a decade or more.

Figures 9–11: The 1436 cloister and interior courtyard at the monastery of San Michele, original to the monastery during the time when Fra Mauro drew the map. Presumably, Fra Mauro lived somewhere on the upper floors of the older section.

Figure 12: Venetian merchant-explorer Marco Polo who traveled though Asia for 25 years in the 13th century with his father and uncle. The written tales of his travels are Fra Mauro's main source of information on the eastern portion of his mappamundi, and Polo is often quoted verbatim, but Fra Mauro does not cite Polo specifically.

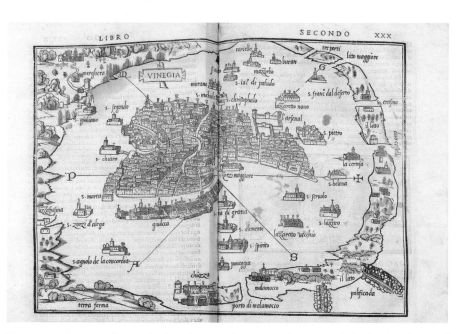

Figure 13: 1528 map of Venice by Benedetto Bordone, printed in Venice for his book *Isolario*. Off the northern edge of Venice are the two islands San Michele and San Christoforo which were later joined to become the cemetery island.

Figure 14 TOP: Earthrise as seen from the moon, the first time a human was able to see and photograph their planet in full. Taken in 1968 by Apollo 8 astronaut William Anders while orbiting the moon. **Figure 15** CENTER: The *Babylonian Map of the World* (500-700 BCE), also called *Imago Mundi*, is considered the oldest world map. **Figure 16** BOTTOM: World map by Eratosthenes (194 BCE)

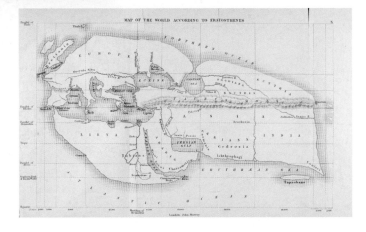

Figure 17: Reconstruction of the 9th century world map by Ibn Sahl al-Balkhī (850-934 CE) who started the Balkhī school of terrestrial mapping. None of the maps from that school survive, but as the reconstruction shows, the map is oriented south, and it was the first world map to do so. Unlike his peers, Fra Mauro also used this orientation. The al-Balkhī map was also the first to make a "schematic" map of the world (beyond the ancient T-O maps) by using thick lines to represent rivers and separate land masses.

Figure 18: This map by al-Istakhri, created about 973, is oriented south and it is drawn as more land than water. Note also that the Mediterranean drips off to the Atlantic Ocean. This and other Islamic maps are known for their artistic beauty as shown here.

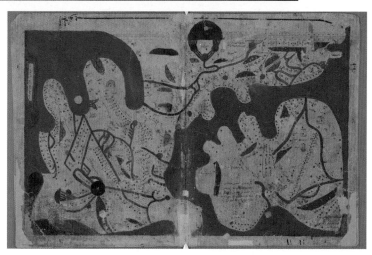

Figure 19 TOP: The Arabic *The Book of Curiosities* contains a world map, a 13th-century copy of the 11th-century original. There is also a round world map in this two-volume book, but the author and cartographer are unknown. This is the first map to have a scale on the side to place geography with mathematical accuracy. All the place names are in Arabic and the cities are important to Muslims.

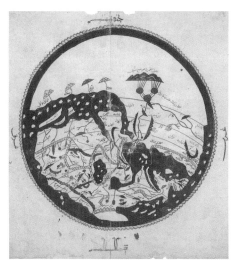

Figure 20 CENTER: A 1456 copy of the Al-Sharīf al-Idrīsī (1154) map of the world in *The Book of Roger.*

Figure 21 CENTER: World map by Ibn al-Wardī (1349) as part of his manuscript *Kharīdat al-ʿAjāʾib wa farīdat al-ghaʾrāib (The Pearl of Wonders and the Uniqueness of Strange Things)*, and copied in the 17th century.

Figure 22: Map of Venice drawn in 1525 by renown Islamic cartographer Piri Reis. Only a fragment of his world map remains but his more local maps, many of which survive, are celebrated for their artistic beauty.

Figure 23: The *Da Ming Hunyi Tu* is a huge Chinese-made world map painted on silk in 1390 with China at the center.

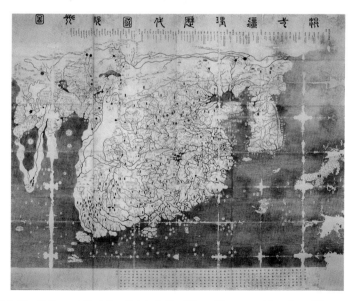

Figure 24: *Honil Gangni Yeokdae Gukdo Ji Do (Map of Integrated Lands and Regions of Historical Countries and Capitals)* is a Korean world map from 1402 often called the Kangnido map. One extant copy is on silk and the other on paper. China is the center but interestingly, Africa is circumnavigable. And this map shows Europe and Africa in line with Eastern trading to and from those continents.

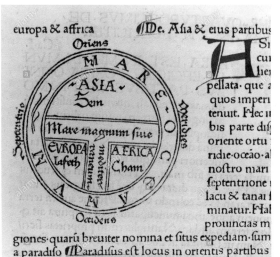

europa & affrica   ¶De. Afia & eius partibus

Oriens

MARE·OC

ASIA

Sem

Septentrio

Meridies

Mare magnum fiue

EUROPA

Iafeth

mediterraneu

AFRICA

Cham

Occidens

giones·quarú breuiter nomina et fitus expediam·fum

a paradifo ¶Paradifus eft locus in orientis partibus

Figure 25: T-O map, drawn for the first time by Isidore of Seville in the early 600s.

Figure 26: The Cotton Map, 11th Century.

Figure 27: The Sawley Map (1190, Great Britain). This map was part of a religious manuscript and measures only 8.5 by 11 inches. It is covered with inscriptions and includes the first depiction of Gog and Magog.

Figure 28 ABOVE: The Ebstorf Map, mid-13th Century, Germany. A large map, 12 feet by 12 feet, drawn on 30 goat skins. Note the head of Jesus at the top and his hands and feet placed accordingly on the map. Also, note there is writing all over this map and in the corners as Fra Mauro's map would be 200 years later. It is also an encyclopedia of its times. **Figure 29** BELOW: Hereford Map, 1300s.

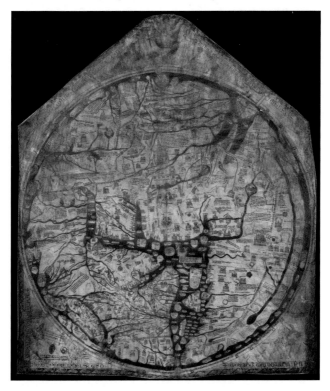

Figure 30: The Vesconte World Map (1321) is a map but it also has the rhumb lines of a typical portolan chart that would be used for navigation.

Figure 31: This map was a part of a nautical atlas which is signed and dated in 1439 by Andrea Bianco. Bianco was a sailor and cartographer and he collaborated with and greatly influenced Fra Mauro.

Figure 32 ABOVE: The Walsperger Map of 1448 (Germany) which makes it contemporaneous with Fra Mauro's map. It is also oriented to the south and has an undulating coastline. In some ways the Walsperger map is a simpler, less cluttered imitation of Fra Mauro's map and echoes the same interest in making a map that is more accurate geographically. **Figure 33** BELOW: Genoese World Map, 1447. This lozenge-shaped world map was clearly made in Genoa, Italy as indicated on the map. It is oriented north. Rumors suggest Columbus may have used this map to sail across the Atlantic, but there was little to guide him.

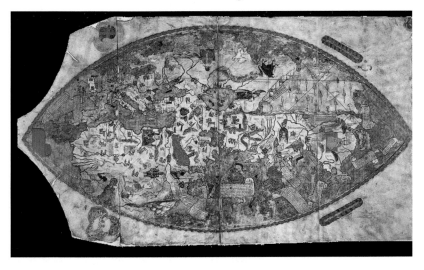

Figure 34: Leardo Mappamundi (one of three versions), about 1450, Venice. Fra Mauro
must have been familiar with these maps since he was working at about the same time.

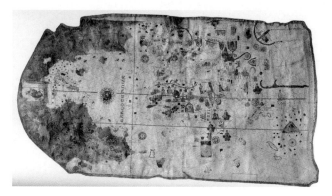

Figure 35: Castillian explorer Juan de la Cosa's 1500 map shows the "New World" as swaths of green land not connected to Asia. De la Cosa had been on two voyages across the Atlantic with Columbus and three other expeditions and was one of the first Westerners to land in South America. He later died there as indigenous peoples resisted Spanish conquerors.

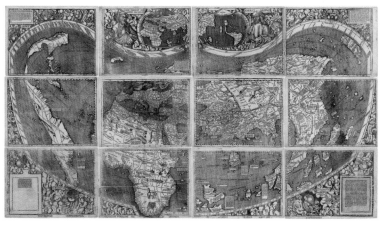

Figure 36: WaldseeMüller Map, 1507, France, first map to include the name "America" and show the Americas as separate from Asia. Often called "America's Birth Certificate," the last extant copy is in the Library of Congress.

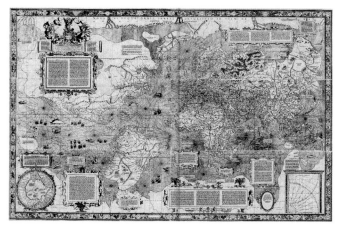

Figure 37: The 1569 Mercator Map with a two hemisphere projection and lines of latitude and longitude. After a large print run, only two copies have survived.

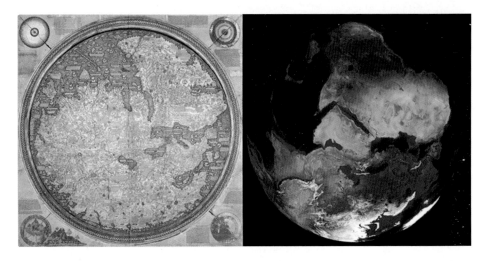

Figures 38–39 ABOVE: Comparison of Fra Mauro's map with a Landsat reconstruction oriented for the same view which shows how much Fra Mauro, who had never seen the earth from space, was geographically correct. **Figure 40** BELOW: Apollo 14 astronaut Edgar D. Mitchell pictured consulting a map while exploring the Fra Mauro formation on the moon in 1971.

THE VENICE OF FRA MAURO

Although Fra Mauro was often living the life of a cloistered hermit, he also engaged with the outside world by collecting rents for the monastery and as he worked on the map with his chosen team. Lucky for him, in the fifteenth century, Venice was a happening place, especially for a cartographer.

From the twelfth century onward, the Republic of Venice was a major player on the international scene because of its trading economy. The city, and thus the republic, had grown in size and influence starting in the 1100s because of its connections with both Byzantium in the East and Europe in the West. This enviable position as a trading superstar was a matter of both geography and economic skill, and Venetians were smart enough and adaptable enough to turn what might have been a disadvantage—a city built on water—into an advantage. They knew boats, how to build them, and row or sail them. Venetians could also expertly navigate the vast lagoon and head out into the Adriatic and the Mediterranean. And they quickly figured out how to form trade routes using both water and land, moving goods from one place to another rather than producing or growing anything and trying to sell it directly. Venice was also not politically or economically hampered by the sort of dogmatic paranoia of city-states such as Florence and Milan. Instead of the one-family rule of those cities, the Venetian government developed as a committee of committees because the original population of the area was scattered across the various islands of the lagoon.[45] Also, all the citizens of Venice were involved in trading, from the rich who paid for the trade boats and ran the warehouses to those who sailed those boats and loaded and unloaded the goods. As such, there was a sense of a collective that was oriented toward economic success for everyone. That community was protected by a government that by and large looked out for the collective. For example, Venice invented quarantine in 1348 to protect Venetians from the incoming plague, and it was the first city to have a Department of Public Health (Magistrato di Sanità).[46] The small clusters of islands that made up the central city and the fact that the city was surrounded by extensive water and was therefore tricky to enter or leave, as well as the complex maze of tiny streets and *campos* that had been cobbled together since the 900s, meant that the internal social fabric of Venice was tightly woven. The rich ran into the poor every day as they walked or boated about, and everyone was somehow involved in trade. Everyone interacted on a personal level through business as well.

Early on, they also had different kind of economy. The fact that there was little land for growing crops or husbanding animals, Venetians had to look outward for opportunities. That vision was realized first in the salt trade. Venice began with its own salt pans, but merchants quickly understood that they could make more money as middlemen. Ships arrived from all over, and the Venetian government soon required that they carry salt as a tax. The government then traded that salt using the many rivers that flowed into the lagoon from the Alps and land routes across northern Italy to monetize salt into the rest of Europe.[47] Over several hundred years, Venetian international trade grew to include timber, fabrics, spices, slaves, and anything else they could move from one place to another and make money. The heyday of the Venetian trade was from the 1200s to the end of the 1400s, covering the Middle Ages, the early modern period, and into the Renaissance. In essence, the Republic of Venice was established and fostered by the invention and devotion to capitalism and a concomitant focus on money. It also had a stable and long-running government that cared about the citizenry. Although that government had been inclusive of all classes of male citizens for a long time, in 1297 the rich decided to band together and exclude from governing power those who were not so rich. With that move, called the Serrata, the formerly democratic government turned into an oligarchy, but still, the middle class, called Cittadini, and the working and destitute poor were of concern to the government. Venice was a city that had civic garbage collection and clean water in all the neighborhoods gathered from rainwater, and citywide lighting to guard against thieves.[48]

It's difficult now, when we view Venice as a quaint honeymoon spot, to realize that this small city was once a nation for hundreds of years, a republic for over 1,000 years which makes it the longest lasting republic in the world, and that it ruled the seas and was an oversized player in international economics. But that very history, not just the art and architecture from those centuries that we gawk at now, is what places Venice in a unique place in the canon of Western civilization. Also, because of its highly successful subsistence on trade, it became one of the most highly regarded intellectual and cultural centers of Europe.[49] Venice witnessed the transformation of all aspects of life from the Middle Ages into the Renaissance and was in great part responsible for those changes.[50] Venice also had an advantage over other European cities because it was, by definition, an international community. Traders came from the East and Europe, and crusaders passed through Venice seeking transport ships and supplies for their travel. The open spaces of Venice, let alone the usually fine weather, meant that all day long every *campo* was filled with all kinds of people chatting with each other, as so many of the art works from that time show. The visitors—including sailors, merchant traders, explorers, crusaders, and those who had traveled to Venice to make money—were happy to talk with Venetians about where they came from and what they had seen because that kind of discourse was normal life in Venice. There were also native Venetians who had their tales of the outside world. Venetians such as Marco Polo and his uncles traveled far and wide in the East, picking up goods to trade as well as information about other countries, continents, and cultures. And some Venetian explorers, such as Niccolò de Conte, wandered about seeing the world, bringing back cross-cultural and geographical knowledge.

Venice was always, and remains, a port city, with all that implies in terms of humans mixing their money, interpersonal interactions, and knowledge. As such, more than any other city in the West, Venice was the melting pot of its times, and that mixture made for a community open to new ideas, a citizenry eager for information, and a place at the forefront of progress.

When Fra Mauro was born in the late fourteenth century, the government and the wealthy citizenry made a move to expand the Republic of Venice beyond the islands of the lagoon. Venetians had never been interested in expansion, never set themselves up to be an empire, and yet the economic burden of various expensive wars with Genoa and others pushed the upper classes to look around for a source of income beyond trade. Before, they had only wanted access to trading ports in the Adriatic, the Mediterranean Sea, and the Black Sea, and they had an accomplished, world-renowned navy to take and hold those ports. But the city was vulnerable to all those land-based cities that had money, power, and land armies. And so, in response to a particular aggression by the Duke of Milan that fueled the idea that owning part of terra firma might fill their coffers, Venice began acquiring parcels of land on the western edge of the lagoon and down the eastern coast of the Adriatic. They also took over the cities of Verona and Padua, incorporating them into the Republic of Venice. The intention was to protect their trade routes up into Europe, but money from these conquered cities would also build a land army. Caught in this cycle, there was constant fighting and jostling between Milan and Venice that lasted decades, and Venice ended up with even more territory. Today that would include the cities of Brescia and Bergamo to the west, the provinces of the Veneto and Friuli, and south to Ravenna and Ferrara. Meanwhile, Venice built up its navy to patrol the Adriatic against pirates and make sure they had open ports for trade all along the Dalmatian Coast. Venetians eventually ruled the large peninsula of Istria in present-day Croatia, the Ionian islands, all of Dalmatia, and bits of Austria. Even when Constantinople fell to the Ottomans in 1453, Venice managed to keep their trading with the city intact, but for decades afterward, this part of the world was awash in constant conflict between Venetians and others. Alliances with other cities or nations went back and forth, over and over, and often the land they had gained soon had to be turned over to someone else. In other words, what might have been temporarily called the Venetain Empire, at its height covering twenty-seven thousand square miles and inhabited by 2.1 million people, was a constantly shape-shifting slippery enterprise with few rewards that cost an enormous amount of money to maintain. Those territorial inflections were due, in part, to Venice's inexperience with land wars and a spotty history of trying to conquer and hold foreign territory, let alone mistakes made with alliances that never worked out.

For a while, the expanded Venetian Republic was divided into three subsections: the Dogado, or the main part of Venice under the direct control or guidance of the Venetian leader called the Doge (for Dux or Duke); the Stato da Mar, which focused on the lagoon and the coastlines of the Adriatic and Mediterranean Seas; and the Stato di Terrafima, which referred to land on the mainland of Italy. Many historians now feel that this land expansion and participation in quibbles on the mainland distracted them from making money and weakened the stability and strength of the republic and the combination was probably the start of the decline of Venice. All these fights and land grabs also wrecked Venice's hold on the spice trade, which was fueling the economy, and other countries such as Portugal and Spain stepped into the breach and started sailing their trade routes, inadvertently initiating what we now call the Age of Discovery and Exploration. That age was not so much about exploration and discovery as it was about seeking wealth by finding new resources and goods to bring back home and sell.[51] At that point, Venice was often in third or fourth place in city rankings because it had diverted its interests to the mainland. Yet it took another three hundred years to bring Venice to its knees and end its glorious run as the longest-standing republic in the world, a title it still holds.

But in the mid-1400s, when Fra Mauro was working on the map, Venice was at its peak, an incredibly powerful political entity, a city with economic dominance, and one of the most influential hubs of culture and intellectual life in western culture.[52] La Serenissima (the Serene One) as Venice called itself, was the second largest city in Europe at the time when only Paris was a rival for the most important and active city in Europe.

LIFE IN THE CITY FOR FRA MAURO

Daily life in Venice in Fra Mauro's time was dominated by the various activities of the shipping trade and the infrastructure to support that trade. The waters around Venice were teeming with all manner of boats—large trade cogs, intermediate-sized cargo boats used to go upriver, personal small rowboats, and of course, gondolas, and to Fra Mauro, all those boats would have been his idea of normal life. The city was also crowded with residents as well as visitors from all over, and he would have been surrounded by the sense of foreignness that comes from outsiders.

Fra Mauro would also have been experienced in getting around Venice. Venice is not one island but is instead made up of 118 tiny islets that are connected by over four hundred bridges. It is, in fact, an archipelago. But before Napoleon arrived in the early 1800s and transformed the cityscape by installing bridges and filling in canals to make foot travel much easier, everyone moved about solely by boat. Certainly, Fra Mauro would have taken a boat from San Michele to the city center, and he would have rowed, or been rowed, around the city as well. It's not a reach to suggest that Fra Mauro would have known how to row the Venetian way, which is standing up, forward-facing, and using a very long oar to propel the boat in the direction one is looking.[53] Just about every Venetian, even now, knows how to row a boat this way. Keep in mind that the lagoon is very shallow—on average five feet (1.5 M) deep—and so falling into the water would not have been particularly dangerous.

Beside all the usual hustle and bustle of a medieval city that Fra Mauro must have encountered, the city was also dotted with various confraternities that were run by citizens. These *scuole* were not part of the government and not part of the nobility. They were organizations for the lower and middle classes, places to meet, and a way to contribute to Venetian society. By the 1400s, when Fra Mauro was walking the streets of Venice, there were over one hundred of these confraternities divided into grand schools (those with lots of money) and lesser schools.

The grand schools were so well-endowed they could fund both elaborately designed and decorated meetinghouses and commission reputable artists to cover the walls.[54] Today the most ostentatious, such as the Scuola San Rocco, are art galleries, displaying some of the finest Venetian art in the world, paintings that have remained in place for hundreds of years. Those *scuole* were always affiliated with a neighborhood church and they built their secular meetinghouses nearby.[55] The associations started as religious and charitable groups, but later these confraternities were sometimes based on occupations such as those who used paints, or cobblers. For the occupation-based *scuole*, their meetinghouses were places to exchange ideas and techniques and Mauro might have visited, or been known, at the Guild of Physicians and Chemists, which included those who used colors in their work such as artists and illustrators. There is no evidence that cartographers had one of those associations, but he might have frequented ones for artists, colorists, or bookmakers. Beyond their role as clubs or associations, the charitable role of these organizations was essential to the social fabric of the city because they provided social services not offered by the government. For example, confraternities built housing for the poor, gave money to those in need, pensions to widows, took care of orphans, and carried the dead to their final resting place. When Napoleon began shutting them down in the early 1800s, along with churches, convents, and monasteries, he ripped apart the nongovernmental networks that made the Venetian Republic operate efficiently and successfully. He also took away the civic pride that came along with the charitable works and the loyalty members felt toward each other. But that was the point. Disbanding these organizations resulted in a loss of identity for members who felt a strong bond with each other because of a shared occupation or those who united over works of charity. Napoleon's goal was to break the back of the social networks that held Venice together, change the look and feel of the city, and make it his own. In doing so, he permanently altered Venetian culture and history, and not in a good way. In any case, even if Fra Mauro didn't frequent the bars and cafés of Venice and engage in friendly gossip and information gathering at confraternities, which were all over town during his lifetime, members of his cartographic team possibly did so.[56]

The monk was also living in a city with an unusual hierarchy of citizenry. For a long time, there was no official class designated as the aristocracy of Venice and no inherited titles, and so the stratification of classes was, by all accounts, just based on money and the date of family arrival in the city, parameters much more flexible and obtuse than the social hierarchy of city-states such as Milan, Florence, and Ferrara which were run by powerful and wealthy families. But when the rich changed the rules in 1297 and decided to designate those from the oldest families as nobles, a hierarchy came into play, but it had some flexibility. Political power in Venice was certainly in the hands of those with money who had decided they had earned the right because a few generations back their ancestors had come to Venice, but it was also possible to be right up next to those nobles if one had made enough money on their own. Fra Mauro certainly interacted with people of all classes as he began to gather information for his map (see chapter 4). He was certainly talking with sailors, explorers, and missionaries. He might have even spoken with the nobility and the Cittadini over properties and when he helped with the survey for the Brenta River project because in Venice nobles ran the government along with the Cittadini who were often civil servants. Fra Mauro would have also spent time in the company of every class of person at the various markets around town if he had been tasked with bringing goods back to San Michele. Given his *converso* status, it is likely that Fra Mauro did commune with others outside of the monastery, and that might have been one of his regular assignments.

Women also had a more visible role in Venice than in other western cities, although women of the nobility were sometimes sent to convents because their families considered them extraneous. The only other option for a noblewoman who couldn't find a husband was to remain inside all day, every day. Those women who were lower ranked than nobles had the run of the city and they often worked for the family business. Fra Mauro would have encountered these women as he walked about town, or at least seen them. Collecting rents suggests that he would have had to deal with all sorts of tenants, including women. But in general, his life would have been predominantly with male company at the monastery, when making the map, and also fulfilling his duties in Venice.

It's impossible to know how much Fra Mauro was influenced by the intellectual life of fifteenth-century Venice. He was working just as Western culture was slipping from medieval times into the early modern period and on into the Renaissance. This must have been an invigorating moment for anyone who was educated, read a great deal, or was open-minded. In Venice, the fifteenth century was abuzz with ideas, inventions, and cultural changes. During that century, Venice initiated the practice of providing a public defender, built the first quarantine hospital to isolate plague victims, was the first government to survey its territories, and began a thriving printing business that flowered into over two hundred printers by the turn of that century.[57] There is no record of Fra Mauro being included in intellectual gatherings, no evidence of him writing philosophical treatises, and no mention of him in any of the history books except as a cartographer. And yet, when drawing the map, he had to make a million informed decisions about geography and culture, which suggests that he was well-read and well-informed. We have no hint that he was part of a group of cartographers who met and talked shop, and no suggestions of his position on the way scientific thought and method were taking over medicine at the University of Padua, which was part of the Venetian Republic. We can only assume, based on what is on the map, that he was up to date, and inspired, by the intellectuals, academics, and writers who were leading Venice into the Renaissance.

For Western culture, this was also a time when critical thinking, as opposed to adherence to the word of God, was taking hold. Venice, a city full of churches, monasteries, and convents, also had a shaky relationship with the pope. The city had historically been ignored by Rome, and so Venetians learned to make their way in a religious sense. When the Catholic Church actively tried to interfere, Venetians resisted. It was not so much the church as a religion that they disliked; it was the pope as an authority figure. Venetians never wanted an outsider to tell them what to do, and they viewed the pope as an outsider trying to push his will on the republic. But devotion to the Catholic faith is still imprinted on Venice. There are now about 135 Catholic churches in Venice and the islands combined, although ten of them have been deconsecrated. One record shows that fifty-seven churches no longer exist,[58] which gives a good estimate of how many churches were around in the fifteenth century—quite simply, a lot. Interestingly the Venetian diarist Marin Sanudo, who kept track of everything daily that happened in the government, wrote that there were 137 churches in Venice in 1493, which is just about the same number as today. That stable number means that while some were destroyed by Napoleon's army, and others crumbled over the centuries, still others have been built. For Fra Mauro, that religious atmosphere was so much a part of life that he must have taken it for granted. It's difficult for us now, in a time of secular dominance, especially in Western culture, to understand what it must have been like to live in a culture with such an overarching mantel of a religious spirit. Monasticism was revived in the 1400s and monasteries became common throughout Europe, especially in Venice and on the various island of the lagoon. Although there is no complete list of the monasteries during that century, there were many monastic orders attached to churches, especially the large ones. There were also about thirty convents in the city housing over two thousand women.[59] Many of these nuns were daughters of aristocrats who were deemed ineligible for marriage, usually because their families had run out of money for a dowry, and so their best alternative was to go to a convent.[60] Today there are still at least five monasteries and four convents that operate in tiny Venice and they accept overnight tourists as a way to finance their orders.[61] It's not unusual to see nuns or monks walking about Venice, especially in the parts away from the tourist mobs. During Fra Mauro's time, then, being a monk in what we now consider Italy, and in Venice in particular, was not unusual at all.[62] As he walked through town or visited other islands, he was never far from a church and his presence in a white cassock wasn't odd in any way. In that sense, Venice was a religious place where the church had a strong presence. Even today, non-Catholics can't ignore the site of churches in every small *campo* and at intersections of the various *calle* and the fact that these many churches define the Venetian cityscape. They also give an identity to each section of the city with their buildings, services, and festivals. Tourists are sometimes required to enter these churches if they want to see the magnificent art that remains in place since the time centuries ago when they were commissioned by the parish clergy who still own them. And surely the church bells that ring out all day long, each with its accent, proclaim how closely Venice has always been entwined with the sacred. Although there were other religions in Venice such as the Greek Orthodoxy, Judaism, and Islam, but Catholicism predominated. Mauro, then, was a member of the predominant religion of the city.

Fra Mauro could not possibly have escaped the hustle and bustle of international trade that was the economic engine of Venice. Leaving the quiet of San Michele, he would be set down on the north end of Venice on a promenade. From there, he could have walked among the labyrinth of *calle* that provided alternate routes to any destination, but those routes would have required temporary planks across canals because there were no permanent bridges like there are today. Or, as a Venetian of the fifteenth century, Fra Mauro probably rowed right up to his destination, tied up his boat, and scrambled out onto a stone landing or set of steps. He might have rowed down the Grand Canal to reach Piazza San Marco (St. Mark's Square), the heart of the city, but that route would mean weaving through the many huge trade ships, cogs, and galleys, that clogged the Grand Canal and docked side by side along the Riva degli Schiavone on the southern edge of the city.[63] Once docked, he could have walked into Piazza San Marco to purchase something at the many stores and stalls that lined the Piazza. Or perhaps he had a destination at some other monastery and could avoid the city center and row gently and quietly through back canals to a more peaceful part of the city. When collecting rents, Fra Mauro presumably went here and there in the city, knocking on doors, waiting for someone on an upper story to push out the ubiquitous green shutters and ask who was there and what they wanted. A man in a white cassock would be the answer.

The rumor mill of Venice might have also brought information to him, or he might have heard about other cartographers and other maps in the shops where he went to buy parchment and velum, paintbrushes, and pigments. But we can only wonder who Fra Mauro spoke with as he began to think about and then construct his map. Fra Mauro could have visited a café of some sort to chat with sailors who had information, or maybe he spoke with learned gentlemen who had been abroad, explorers who had seen things, and foreigners who knew different lands than Venetian traders. In these pursuits, he would have had a feast for the eyes because of the lively and complex business of international trade. He would have heard the sound of many languages, the look of people from all over, and smelled the spices transported to Venice from far away. All these sounds were, of course, underscored by the tolling church bells, the rasp of sails being unfurled and lashed to masts taller than most of the buildings in Venice, and the calls of burly stevedores unloading the great ships. Given the collective nature of the monastery, and the interpersonal intensity of Venice, it would be a surprise if he somehow had spent years on a world map working in an intellectual and conversational vacuum. He was also working with a team, and they, too, would have brought knowledge of cartography, geography, and the history of maps to the table.

Fra Mauro was a man who seems to have left his birthplace only once, to go across the Adriatic to Istria, yet he would eventually construct a map of the world. But the most effective resource for that map was Venice itself, a microcosm of the world of 1450. He certainly must have been inspired as a cartographer by Venice's place as the crossroads of East and West. He was also decidedly lucky, as a soon-to-be mapmaker, that he lived in a place where the world came to him rather than the other way around.

CHAPTER 4

Fra Mauro Makes a Map

*Map construction, no less than writing text, is essentially a
social act, one which involves much more than drawing lines.*
— Alessandro Scafi, *Mapping Eden*, 1999

*When he undertook to compile the mappamundi, the task
Fra Mauro set for himself was encyclopedic in nature.*
— Angelo Cattaneo, *Fra Muaro's Mappamundi and
Fifteenth-Century Venice*, 2011

*This work, created for the perusal of our illustrious signori,
is not as complete as it should be, as it is quite impossible
for the human intellect to comprehend this cosmography
or map of the world in its entirety without some kind of
celestial demonstration; it serves merely to impart a taste
of knowledge rather than to satiate a thirst for it.*
— Fra Mauro, Legend on his mappamundi, 1459

I magine this scenario: One day, after much discussion, a group of
monks clad in white tunics gather in the cloister of a monastery on a
small island in the Venetian Lagoon close to the city of Venice. A decision has been made—the lay priest Fra Mauro is going to head a team
that will make a map of the world. They all know this undertaking will
take years, and involve many calligraphers, artists, and geographers,
and the plan is to make a very large map. Where to set up? They needed
a large room and one that could be occupied for a very long time. The
planners probably didn't worry too much about making sure the door to

this workshop could be locked—who would steal or harm anything in a monastery? Perhaps no one outside the monastery even knew about this project, which also made it safe. Maybe they cleared out a storage space, maybe they took over the refectory, or some large room that was used for communal meetings. And then the project began. Boxes of the ingredients for making inks and tempera paints, calligraphy pens, and brushes would start arriving by boat. A stack of velum pelts would also be offloaded, and there must have been a load of wood for the backing and frames of this proposed gigantic map.

Despite the influence the finished map would wield for the next five and a half centuries, the story of how it was built is vague and little documented. We have no notes from the creator, Fra Mauro, but his notes surely existed—there is a trail of their whereabouts until they vanished. We also have no preliminary drawings and no record of conversations. The story of the map's creation is as shrouded in mystery as its creator. All we have are bits of information that historians have uncovered over the centuries and pieced together. And we have the map, which is teeming with explanations of the good father's thinking and decision-making. But there is scant information about how the map was conceived, organized, and executed. Only some brief mentions show up in the monastery books, but they are telling.

THE FINANCIAL TRAIL

Fra Mauro's map is first mentioned in the monastery books of San Michele on February 8, 1447, when Abbot Maffeo Gherardo wrote that one Dom Benedetto Miani had given twenty-eight gold ducats to help complete a world map being made by Fra Mauro.[1] Gherardo also kept the monastery account books from 1448 to 1499 and they frequently note payments for supplies such as paint and gold leaf for the map. The financial logs also have a line about payment to artist Don Francesca de Cheso, who might have painted some of the scenes outside of the inner circular frame in the four corners, or perhaps some of the illustrations of cities, monuments, and temples on the map; he was reimbursed for expenses in 1457. That year there was another line about money paid for a *scriptor*—that is, a writer or calligrapher—to work on the world map for seventeen days. There is also an additional San Michele register by Gherardo of incoming and outgoing expenses from 1453 to 1464 with many records of purchases to complete a map copy for King Alfonso V of Portugal, but the cartographer is rarely mentioned by name.[2] This particular commission, a copy of the map, might have been long in the making; evidence shows that Infante Dom Pedro of Portugal, the king's oldest son, visited Venice in 1427 and supposedly traveled to San Michele, where a general exchange of Portuguese-Venetian cartography might have been initiated, but no one knows for sure.[3] Payments for this copy continue into 1462, but Fra Mauro had probably died by then. The last mention of the cartographer himself is a causal note of eight ducats given to Fra Mauro in 1459. In other words, there are real notations with real figures that give a thin trail of money received and spent by the monastery to make a grand mappamundi. In the realm of historical research, this is called evidence and these numbers are precious data when little else has survived to tell us how and why this map was made.

WHY DID FRA MAURO MAKE THIS MAP AND WHO WAS THE INTENDED AUDIENCE?

World maps back then were not made for the public, but for a rather small audience of intellectuals, aristocrats, and members of a government. Their purpose was to enlighten those who were interested in the current knowledge of world geography and human activity across the globe. In that sense, a mappamundi was an exercise in gathering information and making it legible to others. In some cases, owning this sort of map meant feeling superior, better informed, and smarter than others. Monasteries were in a perfect position to shoulder such a

project because many of them, like San Michele, were centers of learning
and scholarship and so they were already dedicated to synthesizing and
sharing information.

It appears that this map gained attention even before it was started,
and that may or may not be due to a portolan that Fra Mauro completed
earlier. The lay priest also had done that other cartographic work for the
city redirecting the Brenta River, but we don't know how his reputation
led the monastery to trust him with such a large project. And yet Fra
Mauro was enjoined to make the original mappamundi and two copies.
The original was intended for the Republic of Venice, and that's the map
that has survived and now hangs in the Museo Correr in Venice. He also
made the copy commissioned by King Alfonso V of Portugal, although
there is no trace of it today. A second copy was commissioned by the
Medici family in Florence, which is also lost (see chapter 7).[4] Taken
together, these secondary commissions speak to the fact that word of this
map had spread, and it was considered a must-have by powerful people.
Today, Fra Mauro's map is a matter of curiosity for map aficionados and
cartographers, and those interested in the Age of Exploration, and part
of its allure is the mysterious history of its creator and the absence of
decent documentation of its creation.

For that, we must speculate a bit.

FIFTEENTH CENTURY MAPMAKING MATERIALS

One of Fra Mauro's first decisions when planning this map was the size,
shape, and figuring out the necessary materials to begin. He would have
spent time thinking about what the best surface for drawing the map
might be. Presumably, he had already decided to add all sorts of text, so
he had to choose the perfect surface that could support paintings and
detailed calligraphy. Historically, there were options.

Papyrus is made from the pith of the papyrus plant found in swampy
areas. This is a tall, thin green plant with one skinny stalk topped by a

comical puffball of the same color.[5] Papyrus is found growing naturally in Africa and around the Mediterranean, where it remains a vital component of the Nile River's marsh ecosystem. Ancient Egyptians intentionally grew the plants extensively along the Nile River and throughout the Nile delta, but today it is harder to find. It was a popular and useful material in its day because papyrus was utilized in building boats, making baskets, in recipes for medicine and incense, and for tampons. Ancient Egyptians also liked to chew papyrus stalks like sugarcane. Then someone realized that the outside of the stem could be ripped off and the pith inside sliced into strips. Those strips were then laid next to each other with another layer of strips set on top at a ninety-degree angle. Pounding those two layers together made a flat, durable surface for writing. The disadvantage is that a sheet of papyrus is thick and hard to write on, and it's bumpy. Also, when rolled and unrolled, the strips tend to fall apart. In that sense, papyrus is fragile as a surface for writing or painting. Nevertheless, this invention was the first kind of writing sheet, created in Egypt around 2,400 BCE and as soon as there was a man-made lightweight flat surface, people started to draw maps on it. The oldest geographic map in the world is the *Turin Erotic Papyrus*, now located in the Egyptian Museum in Turin, Italy. It was a regional map made for a practical purpose. Today, only calligraphers and paper artists, that is artisans, work with papyrus since manufactured paper is less fragile and more readily available.

Of course, Fra Mauro could have drawn this map on paper. The invention of paper is probably one of the most universally significant, and longest-lasting, technologies ever invented.[6] Papermaking developed in China somewhere during 200 BCE, although it might have been invented centuries earlier. Cai Lun greatly improved the production process by around 100 CE and by the time the recipe for paper making arrived in Europe in the eleventh century, paper had become a popular medium for writers. Paper is a vegetable product made from plants of some sort—cotton, linen, flax, fand hemp, for example—combined with water and old rags to make a pulp. That pulp is then spread thin and dried flat. The great thing about paper is that it is generally smooth and weighs practically nothing so it can be bound together and carried about, as in a book. But it makes for a fragile substrate to hang on a wall, exposed to sunlight and air pollution in a room. Paper also disintegrates relatively quickly and so it is not a good surface for items intended to last the ages. But paper is an inexpensive writing surface, and so it caught on. Of course, the invention of the printing press in the fifteenth century further amplified the usefulness of paper for communication.

Although papyrus sheets and paper were available to Fra Mauro, he had a better alternative at hand and one that had a good track record for maps. Parchment—a common name for the skins of calves, sheep, and goats as a medium for writing and illustrations has been around since at least the fifth century BCE.[7] Humans had many domesticated animals by that time and once they had eaten the animal's flesh, the skin could be stripped off, dried and treated for use as clothing and writing surfaces. No one knows where the process was first discovered, but by the Middle Ages, the process of turning animal skin into a writing surface was in full production. First, after the animal was gutted, the skin was cut off the skeleton and muscle and then soaked in water to remove leftover dirt and blood. The pelt was then placed in a bath of fermented plants and lime, which made the liquid very alkaline and toxic. This vat was stirred with a long wooden oar for days until the animal's hair easily fell off the skin. The trick was to pull the intact skin out of the toxic bath after the hair was gone but before the raw skin began to disintegrate. Once out of the bath, the skin was stretched on a frame and left to dry. During the air-drying, the skin was also scraped to remove the last bits of hair and to make the surface an even thickness. Often, the final product was touched up with powders and such to make sure the ink would not run. Sometimes the dried pelt was coated with egg whites, lime, and flour to make it bright white and even. While drying, the collagen on the skin also acted like a glue that let the pelt keep its shape, even after being removed from the stretcher.[8]

Fra Mauro was making his map right at time when writers and artists were often using paper instead of parchment because of the print revolution, but he chose to work on parchment. There were probably many reasons for that decision. His map was destined to be very large, and it would have needed many pieces of connected paper. Also, paper can tear, disintegrate, and easily stain. Also, paints and inks often bleed across the plant fibers, which makes for a less well-defined illustration. Parchment, in contrast, is tough, durable, and more like the gessoed canvas used by modern artists. Of course, Fra Mauro could have used sail canvas, which is made from cotton, for his map and sailcloth was readily available in Venice through shipbuilders. At the time, most Venetian artists preferred painting on canvas rather than fresco because of the endless water damage that crawled up the walls of Venetian buildings, so the use of canvas would have been a reasonable option for the surface of the map. Instead, he followed the tradition of other mapmakers and chose durable parchment even though it was much more expensive than paper. Historical reports claim that Fra Mauro's map was made on vellum, which is a fine-grade parchment usually, but not always, made from calfskin.[9] Velum was normally prepared with a rubbing of chalk, pumice, or ground-up bones to make the surface extra smooth. The word *velum* was also used for finely made parchment from other animals, so the map might be on a sheep or goatskin that was well-worked, but this is hard to know. But it makes sense that Fra Mauro chose fine vellum, because it was the best possible substrate for a large map. But, parchment, including velum, can also shrink and buckle over time, which explains why Fra Mauro's team glued the vellum onto a flat piece of wood instead of stretching it like an artist's canvas on a frame with horizontal and vertical bars. They would be working on the map for years and making at least two copies just as large, so they needed a substrate that would be stable over time and keep the dimensions and positions of the geographical features correct.[10]

The circle of vellum used for Fra Mauro's map, which is about 6.5 feet (2 M) in diameter, was several skins sewn together, although no source explains how this might have been done or how the seams would be joined and flattened so that they did not show on the upper side. In any case, the velum was laid down and attached to a large flat circular piece of wood, which is probably several flat sheets glued or nailed together. Today we might call it plywood, but there was no such thing in the Middle Ages. That circular piece was mounted atop another disc of wood that was a bit smaller, and together these discs were meant to fit into the round gilded circular frame equipped with cross pieces of wood. The gold circle was also sized to hide the small gap between the map and the framing square of wood. A metal rod was attached to the wood disks, and that rod slid into the hole in the center of the cross pieces and attached the map to the circular frame.[11] That circular frame, in turn, was set within a larger square of wood, which then created "extra" spaces in the four corners. Fra Mauro used those outside spaces for paintings of his cosmographies, the Garden of Eden, and long explanatory texts.

The velum was probably coated first with white paint to even out the base color, seal the leather, and get it receptive to colors. When he was ready to draw on the map, presumably transferring drawings on paper to the velum, Fra Mauro and his assistants probably used charcoal, which was a prominent drawing medium for artists in the Middle Ages. It's also possible that Fra Mauro had made stencils of some kind for general outlines. Once the team began to fill in the map, they must have used goose, swan, or some other bird quill pens and various brushes made from animal hair, because this map was drawn with tempera paints and colored inks.[12] Tempera is an egg-based, fast-drying paint that the team probably mixed with various pigments that were available in Venice.[13] Venice was well-known for its commercial pigments, and a few decades after Fra Mauro's death, Venetians invented the occupation called *vendecolori* (sellers of colors), people who were specialists in pigment production. Their shops were the first art supply stores in the world.[14]

Fra Mauro would have bought ingredients for colored inks and tempera paints at an apothecary, where people went to buy formulations of chemicals, herbs, and animal parts that were used as medicine, for dying hair, and for dying cloth. Medieval ink was made with various recipes, but they all followed a simple process.[15] For example, a common dark ink, surely the black or brown writing on this map, was made with galls, gumball-sized boluses that appear on tree limbs in reaction to wasp damage.[16] These balls, sometimes referred to as "oak apples," are part of the tree's immune system and they are full of gallic and tannic acids, which are compounds important in the hide-tanning process as well as in ink manufacture. The galls were ground to powder and stirred with water, alcohol, or wine. This mixture then sat for a few days, after which green vitriol, a ferrous sulfate, was added, and the color turned even darker. Green vitriol was made with rusty nails and sulfuric acid, the combination of which produced a greenish salt flake, thus the name. Green vitriol was a precious compound, and it was traded from Spain to other parts of the world, which meant anyone in Venice, the greatest medieval trading nation, would be able to find it in an apothecary. That mixture was then filtered and often a thickening or binding agent, such as gum arabic, which is the gum that flows out of an acacia tree, was added.[17] This recipe produced a nice dark color, and it became even darker on the page as it dried. The thickener also made application very smooth with the use of a fine-point quill. The scribes might have alternatively used a black ink made in the same manner using ground charcoal, which was also a common black for manuscripts in the Middle Ages.

As for other colored inks, there would have been any number of plants, flowers, galls, and gemstones to choose from at the apothecary, but one look at the map shows that the main colors they used are black, blue, off-white, red, and green. The most expensive blue at the time was called *azzurro*, made from crushed lapis lazuli. There is a notation in the monastery books of a payment for this color. It is a vibrant glorious blue. Another possibility is a blue called azurite, made from a mineral that was the blue used most often in the Middle Ages, and it was certainly cheaper than *azzurro*.[18] The off-white must have been a lead white made with the mineral cerussite. Women in Venice plastered their faces with a layer of thick white paste called ceruse, which was also made from lead; this fashion has its good part—it covered up facial scars such as pockmarks from smallpox—but it also slowly entered the bloodstream, rotting teeth, and eventually causing death. Venice produced the most sought after face ceruse, and so the team probably had easy access to this pigment. Lead white also dried quickly, which would have been an advantage on such a complex project. Fra Mauro might have even used this white, often called *bianco di piombo* (white lead powder) for the first coat of the map to lock the velum onto a layer of blank paint.[19] The red ink could have been made from the toxic mineral minium, from insects, or a synthetic vermilion that had been around for centuries.[20] The green was probably made from the patina that naturally builds up on copper, brass, and bronze. The chemical reaction that occurs when these metals are exposed to air makes for a green substance called verdigris. As an ink, this green easily turns to brown over time, which is why leaves and grass are sometimes brown on old maps.[21]

To apply the ink, the mappamundi crew could have used miniver brushes, which were made from the hairs of red squirrels, geese, and other avian quills, which came with a tiny slice at the midpoint to hold ink. Quills were sometimes stripped of their feathers. Or they might have used pens made from reeds or metal styluses. A metal stylus must have been used to create the many white wave patterns all over the map. They are perfectly parallel undulating white lines overlaid on a base of blue water. Someone apparently tied several styluses precisely together to make a new implement that would draw several parallel lines at once. And the inks might have been stored in inkhorns, which were real horns with screw-on lids fashioned to prevent the ink from drying up. The tempera was probably applied with brushes made from pig bristles or squirrel tail hair secured together with a wax thread and shoved into the end of a quill as the handle.[22]

The map took almost a decade to complete, so this ongoing project must have been well integrated into monastery life. Surely, the production went in phases—first the design, then building the structure for the map, followed by attaching the vellum. With a piece that large, and that complex, Fra Mauro must have made many preliminary drawings and studies, but as mentioned previously, all his preparative documents are lost.

That means there are no records of what might have been used to do the initial drawings—charcoal, ink, or chalk—before the ink was applied. Eventually, someone must have picked up ink or tempera and started to fill in the overall layer of water and land. Geographic features such as mountain ridges and rivers must have been added later. Then came towns and castles, all the wild creatures, and the various boats and detritus in the seas. Presumably, the written inscriptions that cover the map came last (see chapter 5), but it's hard to tell. They might have been added as the artists and calligraphers had time and may have been done section by section. Archivists do know that many of the original inscriptions were covered over sometime later with small pieces of parchment glued on top of them. These medieval Post-its show that, over the years, Fra Mauro had changed his mind about certain issues and was correcting the earlier text, like any good scientist.

Just looking at this map in a photograph or in person, the details and size alone belie the nine years of work it took to create. We have no idea if the map was worked on every day or if there were periods when no one touched it. But it's probably much like the production of any large manuscript during the Middle Ages. There must have been lots of concentrated work during the day by a team working together, but no one would be working on it at night because there were no electric lights to illuminate the mappamundi and be able to work on all those details. The paint would dry overnight and be ready for more work the following day

Although the map production had many moving parts and many workers, the cost wasn't exorbitant. Using a comparative analysis of costs, historian Angelo Cattaneo makes the case that this map cost less than a fancy gown constructed for an aristocratic Venetian woman of that period.[23] That kind of dress was made from precious fabrics, had elaborate hand embroidery, and was dotted with gemstones and pearls worth a fortune. Therefore, this mappamundi with its various commissioned copies might have been a money-making operation for the Camaldolese order as well as an exercise in knowledge-gathering.

FRA MAURO'S CARTOGRAPHIC TEAM

Fra Mauro was not working alone on his mappamundi for nine years. The most experienced person on his team was Andrea Bianco, a Venetian sailor and cartographer of great repute.[24] In chapter 2, I wrote about the atlas Bianco completed in 1436, a nautical atlas that also has two world maps at the end, one that follows Ptolemy and one that is a more pictorial mappamundi oriented east. Bianco was an experienced seaman and as map historian Evelyn Edson points out, his name appeared frequently in Venetian state records for his various voyages. Bianco went everywhere that Venice traded, across and around the Adriatic, into the Black Sea, throughout the Mediterranean, out into the Atlantic, and north to France and London. On the London voyage, he was the commander of the ship. Other references to Bianco in the state archives suggest he might have been the commander of the entire Venetian fleet at some point, or just high up the chain of command. The job working with Fra Mauro came after his retirement as a sea captain. His fingerprints are all over this mappamundi, even in some inscriptions that rely on first-person tales; Bianco is, in fact, a good example of how Fra Mauro favored first-person observations as the best and most credible resources for his map.

There were also teams of scribes, calligraphers, copyists, and illustrators employed by San Michele to work on the map. At the time, San Michele had a scriptorium, a workshop of scribes, which was known for excellent work and it was a very successful operation. That scriptorium would have included lay monks like Fra Mauro and boys in training to become scribes. There were thousands of manuscripts to copy, books to translate, and letters to dictate, and so this crew would have been busy with their inks and quills as they were in other monasteries, which were the centers for reproducing and illustrating medieval manuscripts. It's easy to imagine a quiet room with heads bent over podiums or tables, quills flittering across pieces of parchment as the words of the greats were preserved.

There were also painters and artists involved in this project. We know this because the monastery register shows that the court of King Alfonso V paid over thirty ducats for painters to work on the later Portuguese copy of the world map.[25] But still, the cost of this labor was not overwhelming. Cattaneo also analyzed the cost of various labors during those years in Venice, and he concluded that all the workers for this map added up to about the salary of a wet nurse for two years or three years of a cook's work, which isn't much to pay for a map that changed the world.[26] Cattaneo also points out that, although mappamundi were luxury items that were desired by elite families, they were also produced rather cheaply. That idea underscores again why the monastery of San Michele might have been happy to have Fra Mauro spend so much time making and then copying this map.

FRA MAURO'S RESOURCES

Fra Mauro did not conjure up his mappamundi out of thin air. It was based on history, adventure, exploration by others, previous maps that he may have seen or known about, and books. Presumably, most of the written works he looked at would have been in manuscript form, although some might have been printed with static type or with engraved woodblocks. Many of the available written works would have been on parchment, vellum, or even papyrus, and they would have been rolled into scrolls and stored that way, or perhaps cut into separate pages and bound together at the spine with wooden back and front covers for protection. This particular form of book is called a codex. Codices are gigantic in format and stacked high with leaves or folios. They are also heavy, but still easier to deal with and store than a rolled scroll, and no one was carrying scrolls or codices about like we carry books today. Both scrolls and codices also needed lots of room for viewing. They require a pedestal, podium, or table for holding the material and plenty of room to open and move along its length or pages.[27] Fra Mauro had no access to books printed on paper because he was dead before the printing business came to Venice. The printing press, invented in Germany in 1440 by Johannes Gutenberg, only arrived in Italy in 1467, landing in a small town outside of Rome. The first patent for a printing press in Venice was given to another German printer who had moved to town in 1469 carrying a press in two suitcases. Three years later, printing shops began to pop up in Venice, but by then, Fra Mauro was long gone. The absence of the printing press in Europe means all the reading materials that Fra Mauro consulted were hand-drawn manuscripts and codices. There might have been some woodblock printed items, or maybe even some woodblock maps, but those were rare. Because all these materials were inscribed by hand and illustrated by hand, they were precious objects, unlike the way we think of books today, printed in the thousands and millions. Also, medieval manuscripts were read by a certain class of person, someone who could read, was educated in some way and had the curiosity to find, read, and haul them around. Fra Mauro was an educated person and presumably, since he had no role in any university, he was taught to read by people in his family and perhaps educated further in the monastery system, one that valued learning and knowledge.

Fra Mauro's sources for reading material were potentially broad. He would have had access to manuscripts and codices from the renowned library of San Michele, from other monasteries and convents in the Venetian Lagoon area and cities close by, from libraries in Venice, and personal collectors. Angelo Cattaneo has suggested that lending and borrowing written material was very common at that time, which meant Fra Mauro had many opportunities to look at books about maps and geography.[28] Surely, he would have read Ptolemy's *Geography*, which had been translated into Latin in 1406. An analysis of an eighteenth-century list of works in the library of San Michele also suggests that Fra Mauro had access to at least ten of the items explicitly cited on the map at the monastery library.[29] But other manuscripts essential to his work, including books on cosmography, were not close at hand, and so he must have borrowed them.

Fra Mauro's map depended on the descriptive travels of others, and he was more interested in first-person accounts of geographical space or geographical experience than any other type of resource, including philosophical or historical accounts that repeated myths and tales. He considered explorers and sailors the most trustworthy sources for his map because, as historian Edson translates one of his legends, they were imparting geographic information they had seen with their own eyes and heard with their own ears (*"veduta e occhio"*)."[30] He trusted these firsthand accounts more than any other resource, including classical written works. For example, Mauro wrote on the map, about Ptolemy in particular, but this quote also stands for everything on the map: "I therefore declare that in my time, I made every effort to examine the veracity of [Ptolemy's] text by experience, inasmuch I spent many years researching and talking to credible persons who have seen with their own eyes what I faithfully describe above."[31] It's the telling words "with their own eyes" that underscore what was most important to him. He was probably familiar with the writing of Odoric of Pordenone, a Franciscan friar who in the early 1300s made several trips away from Italy to the Balkans, Persia, China, Tibet, India, South Asian islands, among many other places. Odoric's travels were well-known throughout Europe, and his manuscript had been translated into French and Latin. Another friar wrote down the narrative of those travels in colloquial Latin and a copy dated 1350 is now housed in the Bibliothèque nationale de France. The finished mappamundi also shows that Fra Mauro depended heavily on Marco Polo's *Il libro di Marco Polo detto il Milione* (*The Book of Marco Polo*, also called *The Million*) better known as *The Travels of Marco Polo* or *The Book of Miracles*. Marco Polo was Venetian, and part of a famous trading family; he spent twenty-four years traveling with his father and uncle, who had already been to China, along the Silk Road and living in various places in Asia, Southeast Asia, and the Middle East. The book recounts his adventures and observations, but it wasn't written by him. The stories were written down in the late thirteenth century by Rustichello da Pisa, a romance novelist, who was a cellmate of Marco Polo's.

Long past his Silk Road adventures, Polo, as a Venetian aristocrat and
wealthy trader, had financed a warship and he was captured by the rival
Genovese and thrown in jail for a time, where he lived with da Pisa.
Therefore, it's hard to know if the stories and words in the book are
Polo's, or if they were essentially true but a bit exaggerated by Rusticello
da Pisa. That book, a handwritten manuscript composed in a combination of Venetian and French, was a blockbuster, and remains so. It was
translated into at least four languages soon after the first copy appeared
(and into many other languages since then) and was a fourteenth-century
best-seller.[32] It's still read widely today and is considered a masterpiece
of travel literature; in a sense, Marco Polo invented the travel memoir
or journal. It is also a document about the business of trade. The book is
full of details about specific trade goods, their prices, and possible profits,
which makes it a business guide for the Middle Ages. Marco Polo also
came home with gems sewn into his clothes, so that part worked out well,
and he proved himself as a trader. But initially, many people, especially
fellow Venetians, didn't believe these fantastical tales of other lands
and cultures and thought he was making it all up. History has verified,
however, that Marco Polo did travel to all those places, his geography
was surprisingly correct (see chapter 7), and his descriptions of various
peoples and cultures were accurate. But the book contained no maps.[33]
And yet, Marco Polo was once asked to paint a map on the wall of the
Palazzo Ducale (Doge's Palace), the seat of government in Venice, but
this map was later destroyed. Given that Polo was a Venetian, Fra Mauro
would have known that travel manuscript well, and he refers to it all over
the map. He also seems to have relied on descriptions by Polo of various
castles, temples, and monuments in far-flung places to make illustrations
of things he had never seen and that were not depicted in any other
books or codices he might have referenced. That these illustrations are
so accurate is a testament to Marco Polo's detailed descriptions, which
Fra Mauro trusted implicitly. Fra Mauro also used many of the original
place names in their original languages in Polo's book and these, too,
turned out to be accurate.

Fra Mauro also must have known about the writings of the Venetian adventurer Niccolò de' Conte, who set off to see the world in the early 1400s. His trip also lasted twenty-five years and, like the Polos, de' Conte spent extensive time in various places, even hunkering down in the Middle East and learning Arabic and Persian. He also became a Muslim. De' Conte sailed among the Spice Islands, including Java and Sumatra, using local boats, made notes about cultural practices in India, went elephant hunting, and got a tattoo.[34] His travels were written down by a papal secretary in 1444 to atone for de' Conte's conversion to Islam, and it is likely that Fra Mauro knew about his adventure and probably read the manuscript. Fra Mauro might have also had in-person conversations with Niccolò de' Conte in Venice. It seems clear that de' Conte description of gigantic Chinese cargo ships, called junks, informed Fra Mauro's drawings of these ships on the mappamundi.

The priest might have also heard about, even investigated, the tales of other Venetian explorers, such as the travels of Nicolò and Antonio Zen, two brothers who supposedly traveled around the North Sea in the late 1300s.[35] An account of their travels, including maps, was set down by a great-grandson in 1558, so if Fra Mauro heard about these travels, it must have been by word of mouth. His charting of various islands in the North Sea was probably also influenced by Venetian explorer Pietro Querini, who was shipwrecked on the Lofoten islands off Norway in 1431.[36] Surely such an experience was talked about all over town. There was also extensive ongoing exploration by sea during the time the map was made. For example, working for Prince Henry the Navigator of Portugal, the Venetian Alvise Cadamosto sailed down the west coast of Africa in 1455 and 1456. He and his crew "discovered" the Cape Verde islands and found the mouth of the Gambia River. Again, such discoveries must have been bandied about in Venice, especially since the explorer was Venetian. So much of Fra Mauro's map relies upon, and seems to trust, first-person accounts over theories of classical philosophers or religious mythology. This point of view alone sets him apart from other medieval mapmakers.[37]

But still, Fra Mauro considered, or took seriously, other authorities, sometimes contradicting or objecting to their views. For example, the idea and words of various theologians and natural philosophers, and previous classical ideas about the heavens, influenced his take on the cosmos. The Bible was, of course, also such a resource.[38] Angelo Cattaneo cites as influences the teaching of various church fathers such as St. Augustine and St. Paul as well as classical philosophers including Aristotle and Avicenna as resources for Fra Mauro's overarching theoretical approach to this map. He must also have known of the work of great geographers who came before him, including Pomponius Mela, the Roman geographer from the first century CE, Pliny the Elder, also a Roman living and writing in the first century CE, and all the Greek and Islamic geographers mentioned in chapter 2. Historian Piero Falchetta, an expert on the texts of this map, suggests that one need only look at Fra Mauro's writings in the area of the Indian Ocean and see the Arabic names to confirm that he had precise knowledge about these places from Islamic sailors and cartographers, and that he must have seen their portolan charts.[39] Some scholars think that Fra Mauro might also have gleaned information about North Africa by way of a delegation of Ethiopian Christians who came to Florence in 1441 during a council of the Roman Catholic Church that was attempting to unite Eastern and Western Catholic faiths. He might have also received information about Ethiopia from another monk at San Michele who reportedly traveled to Ethiopia and back, or from Coptic monks visiting the Venetian monastery.[40] Those possible resources would explain why his Ethiopia is so accurate—because he received firsthand reports.

On the map, Fra Mauro cites a total of twenty-five different persons, such as philosophers and geographers, as references, sometimes agreeing with them, sometimes not.[41] Stepping back, these legends are like citations in an academic paper today. The modern purpose of citing others is to confirm that the researcher knows the back history of whatever he or she is working on, even if the current author does not agree with the finding that went before. Fra Mauro's mappamundi is a cartographic example of that kind of process laid bare—I know this idea, I can explain it, and this is why my current illustration, comment, or conclusion is accepting or rejecting that idea. Again, we see Fra Mauro's mind working like the mind of a scientist in the twenty-first century.

Another of Fra Mauro's most important resources, especially as a citizen of an established seafaring republic, must have been the many portolan charts circulating before and during his time. His hiring of Andrea Bianco as a collaborator is a case in point. As both a sailor and cartographer, Bianco was one of those in-person sources that Fra Mauro depended on, and Bianco had also drawn portolans and world maps. He had also seen some of the world beyond Venice. But there were other portolans that Fra Mauro surely saw or heard about. Keep in mind that portolans were not precious objects commissioned by the wealthy to adorn their walls. These practical navigational maps were taken aboard ships, passed around, and it is easy to imagine them lying around the Venetian Arsenal, the massive complex that was the center of Venetian shipbuilding. Also, there were other private shipyards in Venice and they, too, were probably stacked with portolans. Many non-Venetian portolans also passed through Venice by way of merchant vessels coming from the Near and the Far East and every port in the Mediterranean. It's also probable that Fra Mauro had seen and studied Islamic portolans, since various themes that show up on those portolans are repeated on European navigational charts.[42]

The other piece of evidence confirming that Fra Mauro utilized portolan charts is that he made at least one himself. A copy of that portolan now sits in the Vatican Library.[43] This copy is a rather large 52 by 30 inch (132.5 x 75 cm) navigational chart drawn on three sheets of unequal-width parchment. It was drawn as a copy of one made in Fra Mauro's workshop. Historian of cartography Roberto Almagià suggested that it must have been overlooked, or was unknown for years, because according to him, it represents the best portolan chart ever drawn.[44] This extant copy of a Fra Mauro portolan seems to have been cut off on the sides. It has the usual neck of an animal at one end, but that neck is apparently not reflective of a natural parchment neck and maybe have been added. Oddly, the chart is oriented south (other portolans are always oriented north), which makes it the only portolan in history with this orientation. It has many geographic features, and scholars believe it was done in the style of the *Catalan Atlas* (see chapter 2). Unlike those charts, it is covered with squiggly blue lines, denoting nameless rivers. What makes this a Fra Mauro map are the legends, which are explanations and opinions that speak to the viewer. Portolans are not usually chatty documents, but this one is, which prefaces the look of his mappamundi. Unfortunately, many of these legends are now illegible. Sentences often run into each other, words are abbreviated, misspelled, or changed around, and the grammar is often wrong, which makes reading them difficult. There are also signs of someone trying to fix some of these errors, but these corrections sit right next to the original errors and may be the fault of the copyist and not as they appeared on the now-lost original.

FRA MAURO'S GEOGRAPHIC DECISIONS

Now that he had studied and immersed himself in the rich geographical knowledge that was available to him, Fra Mauro had to make his first cartographic decision: the orientation of the map. When hanging on the wall should the world be oriented north, south, east, or west? All world maps are now oriented north, but it wasn't always so, and Fra Mauro had to make a major choice before he put ink to vellum. Nautical charts were oriented north because they were trying to portray the sailing routes just as they were orientated with a compass, which points north. As described in chapter 2, most of the medieval mappamundi before this one were oriented east for religious reasons, although there was some variation. He chose to orient this map south, like the many much earlier Islamic world maps, but it is doubtful that he intended to follow Arabic world maps when he probably hadn't had all that much exposure to them. Angelo Catanneo argues that the choice of a southern orientation, or "upside-down" as we call it today, might have come from the influence of Aristotle and his components of high, low, right, and left. But even more compelling, Cattaneo writes, is the fact that Fra Mauro was Venetian, and so his eye was oriented south like a trade ship headed down the Adriatic, into the Mediterranean, and hopefully toward the trade markets of the Indian Ocean, which were south and southeast.[45]

Fra Mauro also had to make decisions about the placement and scale of landmasses, rivers, roads, oceans, and cities, as well as their shape. The division of the world into three continents was normal for him or any European cartographer. So, of course, the landmasses on his map would be divided into three—Asia, Europe, and Africa—but he also found the decision about the division of the continents rather boring. He wrote on the map, "Regarding the divisions of the world, that is, of Asia from Africa and of Europe from Asia, cosmographers and historiographers give various opinions. Of these, one could discuss at length, but because it is boring to dwell on this controversy, I will make a brief note of their opinions and leave the prudent to decide which one they should hold as best." He also called this idea "*material tediosa*" (a tedious matter).[46] But he also was the first cartographer to discuss this issue in the legends on his map.[47] The three continents of Europe, Asia, and Africa were always the cornerstone of the medieval map from the T-O maps onward, but his changes in scale for the three continents were different from those earlier maps. His Europe ends up much smaller than on other mappamundi while his east has grown, making the comparison of these continents closer to reality than other renditions. At the same time, Fra Mauro also emphasized that there were three clear continents by naming them in large gold letters on each continent several times—Europa, Asia, and Africa—but he also uses that same gold lettering for regions.[48] He disagreed with others on the details by which those continents would be divided. He used the Volga River, rather than the Don, as the border between the continents of Asia and Europe because choosing the Don River cuts off too much of Europe and it was too twisty, and no one knew its true source. He also chose the Red Sea (that is, the Arabian Gulf) as the boundary between Asia and Africa rather than the Nile River.[49] One historian points out that even when naming Europe, Fra Mauro made sure that Europe was not depicted as a unified whole but an amalgam of various peoples and cultures.[50] Keep in mind that world maps back then relied on geographical features as boundaries, not necessarily political boundaries, and thus the origins and courses of rivers were very important to cartography because they naturally delineated areas. And so, he wrestled with the decisions of previous cartographers to use various rivers to divide the continents, and he corrected many of their mistakes.[51] To Fra Mauro, all geography was up for discussion, and he did so in the legends (see chapter 5).

In terms of scale, Fra Mauro's map stands out because the size and shape of his continents are closer to reality than any other previous map. His Asia is expanded, which in turn dwarfs Europe. He wrote on the map, "Let people not be surprised that in Europe I have shown cities so small and in Asia so big. Where I have had space I have made the places big; where I have been short of space, I have made them small. Let those who see them bear with me if they do not find them totally satisfactory and fully as they would want."[52] In other words, the decision about how big cities were was simply a practical matter if a cartographer was following the actual size of landmasses. On his map, the Indian Ocean is also upgraded from an enclosed sea to an open ocean, and he expanded the China Sea. Both bodies of water are full of islands, and they are described in a positive light.[53] To Fra Mauro, these eastern seas and oceans were no longer mysteries but knowns; they were certainly well-known to Islamic, Indonesian, Chinese and Japanese traders who sailed from the East to trade with Western countries. Venetians, on the other hand, established trade routes that hugged the Adriatic and Mediterranean and flowed into the Black Sea.[54] Before the circumnavigation of Africa, goods for trade into the West had to be transported across land for the last leg of their journey and into a port city where they could interact with Venetian merchants who took it from there. In other words, it's not as if Venetian sailors were going into the Indian Ocean and bringing back spices.[55] All these trade routes, Fra Mauro's map shows, were international networks, like the interconnecting sea routes where today's cargo ships, for example, leave Hong Kong and arrive at port cities such as New York, Los Angeles, Marseilles, and Rotterdam, where goods are then offloaded and then taken to the interior by trucks and smaller boats. It can still be said, though, that Fra Mauro would have grown up knowing well that the larger picture of international sea trade networks was part of his Venetian DNA. Andrea Cattaneo suggests that his attention to waterways was a product of his life as a Venetian, an identity that comes with an innate interest in water, boating, navigation, and trade routes.[56] According to Cattaneo, Fra Mauro's purpose for this map might have been to draw an accurate map of trade routes by water across the world and how they could be joined into one giant commercial network. This suggestion is underscored by the many legends that address issues such as the rising and falling of tides; his drawing of southern Africa, which indicated ships could pass from the Atlantic into the Indian Sea; emphasis on port cities popular for trade; and the fact that in the legends near certain cities he wrote about what was traded there. "Thus, we have a pioneering cosmographic representation of navigation and the spice trade as perceived in Venice in the mid-fifteenth century," Cattaneo writes.[57] This perspective is underscored by the many ships of various types that bob all over this map, suggesting that with a good enough boat, you could go anywhere, pick up any sort of goods, and bring them back to be sold for money.

Fra Mauro also adds Japan to the map, and this is one of the first times Japan appears on a Western map. And his Africa looks more modern because of the large section on the top left of the continent (oriented north), which prefaces the Gulf of Guinea before Portuguese explorers had managed to go that far south.[58] His West Africa is out of whack from what we know now, but the seeds of the bump on the western top of Africa are there. We also see the bottom tip of Africa curved to the right and aligned with the gold frame that circles the world map. There is also blue water around the bottom of the tip of Africa tip, displaying visually that it was possible to circumnavigate that continent and enter the "Indian Sea." Mauro was surely convinced by reports of Arabic sailors who had gone around the tip east to west and then back to the Indian Ocean. And there were tales of Phoenician sailors passing through the Red Sea and down the east coast of Africa, around the tip, and back up the west coast of Africa into the Mediterranean in 600 BCE. The tip of Africa was called the Cape of Diab back then, and on Fra Mauro's map, a large section of southern Africa is sectioned off from mainland Africa by a large river that runs, orienting the map north, from the east to the west. Some have speculated that the cut-off section, which is quite large, is meant to be the island of Madagascar but it fits better as just the tip of Africa.[59]

Along with rewriting Africa as a continent, Fra Mauro's most revolutionary changes to medieval geography came with his drawings of the Indian Ocean and its landmasses and islands. Marco Polo said there were 12,700 islands in the Indian "Sea," but Fra Mauro depicts even more. These islands were, of course, of great interest because they were the source of the many spices that fueled Venetian trade. Those islands include Borneo, Java, and Sri Lanka, and many smaller places such as the line of small islands close to what might be Sri Lanka that appear as some kind of barrier.

In another lesser known part of the world, the Far North (of Europe), Fra Mauro draws a very large Scandinavia, but Greenland is absent. He wasn't sure what to call Iceland, and so variations of that name show up twice in Scandinavia and another time on an island in the North Atlantic. He represents Scotland and England as one landmass but contradicted himself on this in the text where he wrote that Scotland was separated off by the natural geography of mountains and water.[60]

All the land surfaces of the map are covered with the topography of mountains, mountain ridges, lakes, wiggling rivers, human-built cities, temples, churches, monuments, streets, and endless place names in various languages. Because the presentation of these features is generally uniform in format and color across the map, and they only vary in size for emphasis, these drawings hold the map together and make it visually pleasing.[61] Some castles, temples, and monuments were drawn with great care and accuracy and are small masterpieces. For example, the Masjid Al Haram Islamic sanctuary in the city of Mecca is picture-perfect, and yet Fra Mauro had never seen that structure in person.[62]

Other geographic decisions show that Fra Mauro was moving a bit away from religious tradition to reality—he reduced the size of the Holy Land to its actual scale, there is no drawing of the parting of the waters, and Noah's ark is just a little house atop a mountain. In other words, these Christian icons are there, but downplayed, and his reasoning for those decisions are excused in long legends where they are relegated to the historical past unlike previous Western mappamundi where these places are presented as real and on-going.[63] Most significantly, Fra Mauro did not place the city of Jerusalem at the center of the map, and yet he decorated Jerusalem with an illustration of an elaborate Holy Sepulcher, giving the city presence and significance while placing the city more accurately in a geographic sense.[64] Fra Mauro's Jerusalem is a small red dot with a gold star, set where Jerusalem is geographically, at the eastern end of the Mediterranean (if you reorient the map north so it looks more familiar) in present-day Israel. Fra Mauro hedged his decision in this way on the map: "Jerusalem is indeed the center of the inhabited world latitudinally, though longitudinally it is somewhat to the west, since the western portion is more thickly populated by reason of Europe, therefore Jerusalem is also the center longitudinally if we regard not empty space but the density of population." He was suggesting the viewer could just shift their perspective a bit and see that Jerusalem is not at the center geographically, but it is, if you counted heads and squinted you could see it as the center of the world. So many previous Christian mappamundi had Jerusalem at the center that by the time Fra Mauro was working on his map, this placement could be called atypical, making his cartographic decision seem almost like blasphemy. But Andrea Bianco had already moved Jerusalem out of the center of the world map, and several other earlier maps, ones made before the Crusades, didn't have Jerusalem at the center either.[65] Interestingly, the same symbolic red dot with a gold star, albeit a bit larger than the dot for Jerusalem, is set around the circumference of this map at intervals. These starred dots might echo the rose compasses placed at the edge of maps in intervals on portolan charts, or they could be anchors for the cardinal directions of north, south, east, and west and their cousins such as northeast and southwest. Or, in typical medieval fashion, they might just indicate the end of the habitable world.

A major visual impact of this map is the many long texts of explanation (see chapter 5). Another visual treat is the expanse of glorious blue, representing waters full of ships of eight different kinds, local and international, complimented by big fish jumping out of the oceans, barrels and big boxes bobbing about, and even a ship's ladder floating around in between the many sets of undulating white lines, representing waves. Again, these repeated figures bring a sense of order to a very crowded Earth and also emphasize the importance of trade.

Fra Mauro had a new way to indicate the end of human habitation—strings of small islands that arch along the top half of the map at its very edge. The same type of island blobs is placed in what we now call the North Sea and at the edge of China and they curve under what we now call the Indian Ocean, which others thought was a sea. There is also a straight string of them that seems to form a wall between the tip of Africa and the islands of Indonesia. In other words, these strings of islands are not consistent, and they do not circle the whole map. Instead, their placement jibes with what during the Middle Ages was considered the unknown and scary world. Some exploration had happened in all those directions, but the information had not been integrated by cartographers, and so on this map, the islands might be gateposts warning people of unknown danger.

Overall, this map is vibrating with changing geography and human agency. This might be a world that God created, but it was also a human-modified space.

WHERE ARE ALL THE MONSTERS?

At the same time, Fra Mauro's map lacks most of the usual weird creatures and wild humans who might suck your blood, which is usually standard for maps of this age. Fra Mauro was on the fence about this threat. He did not promote the idea that hidden in the back of beyond might be a land full of beasts and one-legged hopping humans, nor did he dismiss that idea. He just left the unknown as wavy lines with adjacent sentinel islets, letting you know that no one knows what lies beyond. But still, he held to some traditional decisions about Christian myths as he put a new spin on those traditions.

For example, like other Christian maps, Fra Mauro also included a bit on the mythical priest and king Prester John (and see chapter 6).[66] The tale of Prester John, a story of a Western priest who somehow landed in India and started converting people to Christianity and making himself a kingdom began in the twelfth century and continued for centuries. Then, Prester John supposedly landed in Ethiopia, again as the ruler of a great Christian nation. No one ever met Prester John, and except for one fabricated letter, no one ever heard from him, yet his legend lived on, and he appeared regularly on medieval maps. Prester John shows up on Fra Mauro's map as well in Central Africa as the master of 120 kingdoms and fighting force of naked men even though there is no confirmation that Prester John ever existed and no evidence that anyone with that name ever stepped foot in Africa.[67]

In the same way, the few monsters and mythical beasts and monstrous people that appear on this map are referenced but then given caveats in the legends. The map sports the traditional medieval barbarians Gog and Magog (and see chapter 6). And yet Fra Mauro writes in the legends that Gog and Magog were probably known tribes and not monsters or barbarians. At the same time, he puts them behind a locked gate while dismissing the idea that Alexander the Great locked them in. Instead, he felt these tribes were held back by natural geography.[68] He also wrote on the map, "Because few people believe in these things, I have made no reference to them here," and "I have found not one person who could give me knowledge of what I have found written; wherefore not knowing any more than I can confirm, I leave to those who are curious to seek to understand such novelty." In other words, there is no proof as far as he is concerned that these imaginary creatures exist.[69] Angelo Cattaneo believes that Fra Mauro was sensitive to the public and knew that belief in these ancient scary figures was fading.[70] Piero Falchetta states, "Far from indifferent to the reception his work might receive, Fra Mauro sets about reassuring the reader as to the care he has put into drawing up the map, the correctness of the decisions he has made, and the credibility of the sources he has used. . . . In short, he is clearly concerned to demonstrate the reliability and the efficacy of the map."[71]

FRA MAURO'S DECISIONS ABOUT THE COSMOS

One of the most unusual, and progressive, decisions that Fra Mauro made was to place his ideas and illustrations about the cosmos outside of the world on the square of external wood that enclosed the round map and its circular gold frame. Take a square piece of paper and then draw a big circle in the middle with a circumference that spreads out within a few millimeters of the edge of the paper. Putting a circle within a square leaves four empty corners, and that's where Fra Mauro placed his cosmology. There are four paintings, one in each corner, with corresponding legends that explain his thoughts about how the universe was formed and how it looks now. These four illustrations include the solar system in the top left, the four elements in the top right, the globe of the Earth with both poles and an equator in the bottom right, and paradise, or the Garden of Eden, on the bottom left.

To include the planets seems like a no-brainer since they were frequently on other mappamundi and geography was so entwined with astronomy in the Middle Ages. This diagram (upper left) is a series of ten concentric circles, all labeled. The first eleven are the planets and the inner four represent the elements, making fifteen rings in all.[72] The diagram shades from the outside gold starred deep blue of the *cielo empirio*, the vast sky of the universe, inward through lighter shades of blue to white with corresponding names for the planets, to an inner section of four rings that represent the humors of fire, air, water, and earth as the center sphere. It's also no surprise that the first humors red ring of fire sits next to the ring of the moon, a nod to the fact that the moon influences the humors.[73] The accompanying legend includes the estimated distances of each planet from Earth and their planetary diameters.

The figure at the top right is a blowup of the inner circles of the diagram on the top left—that is, this figure represents the sublunar world with Earth at the center, and the water, air, and fire that surround Earth as it was pictured in medieval times. The outermost ring, bright blue with wisps of white, is supposed to represent the great waters that cosmologists thought surrounded the whole earth. The accompanying legends explain how the moon influences the tides (an important issue for Venetians)[74] and how land—the earth—might rise above the great waters so that land creatures could inhabit the world.

The schematic on the bottom right is another drawing of Earth, this time as a globe suspended with half of it all water and the other half full of continents. This figure acknowledges that Earth is a sphere and that it is suspended in space. It also has an astronomical chart, but the text is faded or damaged and impossible to decipher. Fra Mauro also wrote in the legends nearby about the texture of the earth using the adjective *porosity*, or concavity, and claimed that landmasses appear because that part of the earth must be more porous and buoyant and so it rises above the all-encompassing waters. This globe is also marked by various latitudinal zones and the accompanying text that speaks of temperate lands where people can manage to live. Where people could live was a major geographic controversy during that time, and so it's not surprising that Fra Mauro took on this issue and drew his explanation and commented on it in the legends.[75] Interestingly, this floating globe of the earth is oriented north, like a nautical chart, and in contrast to the mappamundi that it decorates. As his figure on the bottom right outside the map suggests, half his globe is covered in water and humans should not go there.

Even more interesting, Fra Mauro also put paradise, or the Garden of Eden, outside the map proper, on the lower left. Booting paradise off the map proper was a first for medieval mappamundi; heretofore it was always placed right on the map, usually in the East.[76] This was a heretical decision for a medieval cartographer, and one that points to Fra Mauro's goal to move his map closer to reality and a bit away from myth or faith. In all other Christian, that is Western, mappamundi there is some version of earthy paradise not only on the map proper but it is also usually glorified, made larger than the scale of other drawings. As the name suggests, paradise is supposed to be heaven on earth, a perfect place for humans that was created by God as a peaceful sanctuary where everyone could have everything they ever needed. When Adam and Eve defied God's command and ate the apple, they were banished from this haven and for their sins were sent into the real world to face all sorts of obstacles and hardships. Paradise is also the headwaters of the four rivers that pour into and nourish the world. These four rivers have been interpreted in various ways, including the idea that they represent specific rivers such as the Nile, or possibly four continents. Nonetheless, medieval paradise is always shown with four rivers and those rivers are deemed essential for life on Earth. Fra Mauro's paradise is a walled sanctuary with three towers atop a rocky landscape that represents the world. The Garden of Eden part is placed to the left of the outcrop of rocks, and it contains trees and bushes and a naked Adam and Eve in conference with a fully clothed and haloed God standing by a pool of holy water that is the origin of all four rivers. The Tree of Life, an apple tree that will end up providing the source of the banishment of Adam and Eve, is growing between them, and as a warning, a snake is wrapped around the tree. On the right side of paradise stands an angel guarding the entry door and overseeing the four rivers as they seep under the wall and spill over into the adjacent drawing meant to represent the earth in miniature by way of a large mountain, various hills, and a smattering of trees.

Scholars have also suggested, with good evidence, that this image of paradise was not done by Fra Mauro or anyone else who worked on the body of the map, but by a member of the famous Bellini family, a group that included some of the finest and best-known artists Venice ever produced.[77] Jacopo Bellini was the patriarch. Working during the middle of the fifteenth century, Jacopo was well-known as the best painter in Venice and today he is considered one of the major figures that initiated the Renaissance. The drawings and paintings that have survived show a master of figures, color, and perspective, and an innovator of style. A book of his detailed and magnificent drawings is at the Louvre and his *Annunciation*, a study of how to bring rich gold fabrics to life—as if Mary and the archangel were Venetian aristocrats dressed in their finest—has hung in the Church of Sant'Alessandro in the city of Brecscia for almost six hundred years. Jacopo also trained his sons, Gentile Bellini and Giovanni Bellini. Gentile is less well-known, but his work was prized by the Venetian Republic, and he did portraits of doges and sultans. His masterpiece of a procession in Piazza San Marco, commissioned in 1496 by the confraternity Scuola Grande San Giovanni Evangelista, takes up one full museum wall and is a masterpiece of detail. He also painted a scene titled *Miracle of the Cross at the Bridge of San Lorenzo* for the same confraternity. That painting depicts crowds along a Venetian canal and hanging over a bridge with people diving into the water searching for a bit of the True Cross, which had supposedly fallen into the water; it is one of the most iconic works of Venetian art. These canvasses hang in the Gallerie dell'Accademia (Academia Gallery of Venice). The other brother, Giovanni Bellini, was also a master and his *Annunciation*, housed in the Academia Gallery, shows a Venetian house with a redheaded archangel wearing a flowing gray tunic floating toward Mary. The composition and the execution of this *Annunciation* takes your breath away. It was this hardworking and very talented family, and this workshop, that trained a cousin named Leonardo Bellini, who also had a reputation for fine works and miniatures, and he was likely the artist hired to paint Fra Mauro's paradise.

If one believes that paradise is both real as well as allegorical, these four drawings and their accompanying texts provide Fra Mauro's take on the cosmos.[78] Including paradise in his cosmos suggests he was following the view of St. Thomas Aquinas that the Garden of Eden was both mythical and real, both "earthly and spiritual," and so worthy of being depicted as part of the larger universe beyond Earth.[79] Fra Mauro says as much in the accompanying legend: "The Paradise of Delights does not only have a spiritual meaning; it is also a real place on Earth." But he also notes that this sacred place is "far from human settlements and knowledge,"[80] confusing the matter even more.

Taken together, the four drawings in the corners are not unusual for the age, and they present Fra Mauro's fundamental cosmology, but the fact that paradise is not directly on the map is significant. If he had left it off the piece completely, that would have been even more significant. We might then suggest that Fra Mauro was rejecting religion for a more scientific view of the universe and human history. But as a medieval monk, of course he had strong religious beliefs, and they are shown on the map. After all, placing paradise outside of the map is not a signal that he did not believe in paradise, or that he was such a man of science that he rejected the story of Adam and Eve, but that he joined those who felt the Garden of Eden crossed both worlds, the real and the spiritual.

In essence, Fra Mauro knew the world was progressing, becoming better explored, experienced, and known, and that facts were replacing old ideas. Right there on the map, we see some of the same allegorical figures as previous Western maps, but he also presents his ruminations about the state of the known world and how accurate or inaccurate previous maps had been. Historian of mappamundi Evelyn Edson states, "Here the picture is conventional, while the text is not."[81] That juxtaposition underscores the flexion point between the Middle Ages and the Renaissance and on into the Age of Enlightenment and the birth of science.

Fra Mauro also did not hesitate to question himself about any of the decisions he made, and this humble attitude is all over the map. He repeatedly acknowledges that information about the world was improving, and that meant what he might draw one day could change the next. He saw the representation of geography as something flexible, ready to be altered through new information and discoveries. He explained in many legends why he used more up-to-date labels and how place names and various cultures had changed, and why. In that attitude, Fra Mauro shows again that he was stepping across the threshold of science and the scientific method by incorporating and gaining knowledge by trial and error, hypothesis testing and retesting, and changing his view when necessary. Fra Mauro was a scientist at heart.

FRA MAURO'S MAPPAMUNDI

Fra Mauro and his team worked on the mappamundi from 1450 to 1459, when it was framed and hung in San Michele after his death. Different parts of the map were surely done at various times, and it was an ongoing project for all those years.[82] It's not clear if Fra Mauro was relieved of his other duties around the monastery as the map was created. Nonetheless, the map is dated August 20, 1460, carved in wood on the back by someone else since he had already died. Those nine years of working on the map included the initial and ongoing research, gathering information from sailors and merchants, purchasing supplies, setting up the workroom, drafts of what the map might look like, and managing a team that changed over the years. After it was finished, the map was protected by a wooden cabinet and placed in a room next to the sacristy of the monastery church of San Michele called, appropriately, *sala del mappamundi* (room of the world map). Then it was moved out of the church and into the library of San Michele, where it was mounted on the wall and hard to see. Historian Cattaneo suggests that the lack of light in the library cemented this map's perceived function as something to look at, a visual experience, rather than as a document to be read. In any case, no one could have read all those tiny inscriptions without having decent natural or candlelight and being very close to the map.[83] There the map stayed, relatively forgotten for 350 years, until it was brought to wider notice when another Camaldolese monk of San Michele, Placido Zurla, wrote a study of the map and published it in 1806 (see chapter 7). [84] Then, Napoleon shuttered the monastery in 1810 and all the manuscripts from the library were shipped somewhere else or lost. In some sort of negotiation, Jacopo Morelli, librarian of Biblioteca Nazionale Marciana, made sure 264 printed volumes and 12 manuscript works from San Michele were sent to the civic library, and that deal included this map. Ownership by the library made sense because that building was, and is, a bastion of ancient manuscripts and maps. Fra Mauro's map was handed over to Morelli on July 14, 1811, and it was hung in the Hall of the Sansovino in the library. The following year it was transferred to the Palazzo Ducale, the pink-and-white seat of government across the way from the library. Inside the Palazzo Ducale, it was moved around from the Sala del Maggior Consiglio, the Hall of the Great Council and a place of honor, to the much smaller Sala dello Scudo alongside other big painted maps. Fra Mauro's map stayed at the Palazzo Ducale until 1924, when it was transferred back to the library. The map left Venice only once in 560 years, and that journey was a quick trip to a continent that didn't even appear on the map—Australia. This medieval Venetian map was the centerpiece of an exhibit on world maps at the National Library of Australia in Canberra in 2014. [85]

Today Fra Mauro's map hangs on the second floor of the Museo Correr in Piazza San Marco in Venice, although it is owned by the Biblioteca Nazionale Marciana. Since these two buildings are physically and intellectually connected, that placement makes sense if the library wants the map on public exhibition. Recently, after a long renovation, when it was hidden from view, the map was rehung for public viewing in April 2022 as part of a new permanent exhibit of important world maps from the library's collection. It can now be found after walking to the far end of the museum to a special room accessed through the Grand Reading Room of Sansovino. The map hangs alone in the final dark room of the exhibit, in an obvious place of honor. It's protected by a new black-framed glass-fronted case and indirect lighting that occasionally fades on and off to save the map from deterioration. The sheer size of the map, its textual and geographical complexity, and its artistic beauty is arresting, even startling. In that setting, Fra Mauro's map has finally found its best home, where it is being treated like an irreplaceable work of fine art of great value. [86]

CHAPTER 5

What Fra Mauro Wanted to
Tell Us about Geography

*This work, created as an act of homage to this most illustrious
Seignory, is not as complete as it should be because it is not
possible for the human intellect, without the help of some higher
demonstration, to verify completely this cosmography or that
mappamundi; from these, one gets what is more like a sample
of information rather than a full satisfaction of one's desires.*
                    —Fra Mauro, inscription on the map, 1459

*Although Fra Mauro tried to work within the
conventional format of the mappamundi, his words
explode its certainties and leave questions to be
resolved by future explorers and cartographers.*
          —Evelyn Edson, *The World Map 1300–1492*, 2007

*Fra Mauro's cartographic work is the only major
map to entertain—within its very legends—
an engaging dialog with its reader.*
                    —Alessandro Scafi, *Mapping Paradise:
                    A History of Heaven on Earth*, 2006

*"And where is Venice?" the senator asked immediately.
When brother Mauro showed him, he was stunned, and
demanded to know why he had made her so small, whereupon
the poor friar launched into a long scientific explanation
that boiled down to the fact that Venice was just a tiny
spot on the entire globe. "Then shrink the globe," exclaimed
the senator indignantly, "And make Venice bigger!"*
                    —Alvise Zorzi, *Venice 697–1797:
                    A City, A Republic, An Empire*, 1999

Medieval mappamundi have been called the encyclopedias of their age. The idea that a map can be a book of knowledge comes not from the drawings of landmasses and oceans, but from the fount of information and the geography on these maps. Fra Mauro's map is the best example of such a cartographically based encyclopedia not only because of its orientation and expanded view of the world but also because it is so thoroughly referenced.[1] The map seems to have scribbles all over it, but those scribbles point to Fra Mauro's thinking about where he placed landmasses, cities, rivers, and people. Overall, it's hard to imagine a map more textually explained than this one.

Until the Renaissance, most cartographers traditionally allowed their maps to stand without many words beyond place names and topographical features, but that style was not enough for this monk. He had amassed a great amount of geographical and cultural information and then wanted to translate all that onto a limited surface, somehow to superimpose all that knowledge in a way that was consumable by the viewer. And so, he chose to do that with words. Presumably, Fra Mauro realized that maps were interpretive documents subject to the whims of the cartographer, and so it looks like he wanted to tell the viewer not only what they were looking at, but why he chose the positioning and representations that he did. The inscriptions act as bibliographical sources, giving credence to his decisions. At every line, we know what Fra Mauro knew, where he got his information, and sometimes why he chose one position over another.

Those inscriptions that refer to geography—that is, water and landscapes—are the focus of this chapter, while the inscriptions that address cultural oddities and human behavior are the focus of chapter 6. In general, these inscriptions act as references since Fra Mauro cites who told him the information or where he read it. And so, these texts make the map much more of an encyclopedia than a geographical representation. They also make this map the first concrete step toward scientific mapmaking, a process that relies more on direct observation supported by several in-person accounts rather than stories handed down through generations. As Fra Mauro's map pushed mapmaking forward into a scientific process, it was also the cultural Rosetta Stone between the Middle Ages and the Renaissance. Specifically, Fra Mauro took issue with many established traditions that had carried on from classical times. For example, on the map, he textually questioned why Ptolemy had no knowledge of the Black Sea and pointed out that the classical Greek cartographer knew nothing about northern China or Persia. But, as historian Piero Falchetta explains, Mauro was, in general, following Ptolemy's dictate that geographic knowledge changes over time, and therefore we must expect maps to change.[2] That, too, is a modern scientific approach—perspectives change as knowledge accumulates. Fra Mauro subjected every inch of this map to critical and extensive evaluation; in the inscriptions, we see how his mind worked, and how important it was for him to substantiate his decisions. Even now, the map is a running dialogue between the cartographer and the viewer; he is endlessly explaining how he came to put down a name, topographical feature, or whatever it might be on a particular place on the map.

There are more than 3,000 notes and place names, amounting to 115,000 characters, which is surely more inscriptions than any map that came before or after this one.[3] There are 2,500 toponyms for provinces, cities, watercourses, mountain ridges, and such. On top of those names, there are 300 legends, that is, paragraphs or short bits that are teaching moments written in the third person that give textual support for Fra Mauro's decisions. Unusually, 34 legends are written in first person, and 80 of them use the pronoun *Io*, which means *I* in Veneziano and Italian. Reading these legends feel as if he were standing there talking to the viewer.[4] Those remarks vary in length, from the several long cosmological notes (see chapter 4) and twenty-eight short notes that begin with "*nota che*," or "take note," as a heads-up. There are also some special inscriptions, words on cartouches, which are illustrations meant to look like scrolls that make these legends stand out as important, and many of those are written in first person as well. Some of these remarks have been taken from various manuscripts, often classical ones written by others, and he usually gives the original author credit. To cartographic historian Angelo Cattaneo, these notes seem to "often be used in an adversarial sense with respect to other authors that the Camaldolese friar was criticizing, correcting, and contradicting."[5] Cattaneo also suggests these confronting notes are sometimes aimed at the current viewer of the map, as if the cartographer and viewer were in an ongoing discussion or argument. In that sense, they are interactive rather than simple explanations. By that mode of discussion, Fra Mauro elevates the discourse and places himself in the position of an authority. At the same time, he is asking the viewer to evaluate his texts and decide for themselves, especially when he presents contradictory information compared with other authorities. He is using this map as a learning experience for his audience. "In this sense," Cattaneo writes, "The *mappamundi* is among the first texts in which the readers are made aware of the critical choices made by the author during the composition of the work."[6] But often, Fra Mauro also humbles or excuses himself with words: "Though I have been most diligent in trying to put all the coastlines of this sea [the Mediterranean] in accordance with the most accurate map that I possess, those who are experts should not take it amiss if I am not always consistent. Because it is not possible to put everything accurately."[7] Here we have Fra Mauro talking directly to us, airing his feelings about how hard it is to make cartographic decisions, especially ones that might not be accepted or appreciated by those in the know, and then leaving us to decide whether we accept his logic or not.[8] The map itself is a record of the information he gathered and how he thought through the material as he decided to use, alter, or omit some piece of acquired or traditional knowledge.

These inscriptions also say much about Fra Mauro, Venice, Europe, and the cultural world in which he was working. Like all maps, the Fra Mauro map is an artifact of its time. The original texts pin, for medieval times, what geographers and explorers knew as fact and what was much more vague or unknown. But this map also honors the dynamic history of mapmaking that encourages constant change in the knowledge of geography as well as points of view interpreting that geography.[9] Fra

Mauro was doing his best as an academic during this period leading up to the Scientific Revolution to dissect and reference every decision he had made about every single mark on the vellum.

The verbiage on this particular mappamundi is also done in a conversational, homey, style. If all the inscriptions, the scrolls with long texts, the notes by cities and rivers, and the comments about the cosmos, could speak, this map would be a cacophony of voices, a veritable blizzard of words. The overarching choice to add all those texts, to note every bit of geography, was perhaps Fra Mauro's most significant choice. He could have set those words into a manuscript or book and set it alongside the map for those who might be interested in how he arrived at this or that feature. Instead, he chose to clutter the map with all those words. That choice was revolutionary and effective, and one of the main reasons this map is so important in the history of mapmaking and the history of how humans perceive their world. It should also be noted that in contrast to all those words, there are no human figures anywhere except in the Garden of Eden, which is outside the map.[10]

Out of the three thousand or so bits of writing, most of them relate to place names, but about two hundred are, in the words of Fra Mauro historian Angelo Cattaneo, "A grand treatise of cosmography including descriptions of places and people, commercial geography, history, navigation, and expansion, and, finally, questions that would today be defined as methodological."[11] Fra Mauro knew exactly what he was doing, and that it would be difficult for the viewer. On one of the cartouches he wrote, "If someone finds incredible certain of the previously unheard-of things which I have noted above, he should not submit them to the judgment of his own reason but rather list them among the secrets of Nature."[12] In other words, Fra Mauro knows that some of the viewers would not be intellectual giants, and certainly, most of them were not as educated as he was about geography, so he is also asking for the viewer's trust. Right on the map, he says the viewer should rely on the experts, such as himself, when they don't understand something. The reader is essentially being called upon to bear witness to Fra Mauro's decisions, depictions, and explanations, and this is the first time on any map that a viewer has been enticed into the very process by which a cartographer comes to geographic conclusions.[13] For example, Fra Mauro explains the rise and fall of the Nile waters just as Pliny does, which means he mixes astronomy and astrology with religious celebrations: "The Nile begins to rise at the first moon after the summer solstice, when the sun is entering Cancer; it swells and overflows in Leo; stops in Virgo and subsides in Libra. That is, between when it begins to rise and then stops and falls, from mid-June to the Feast of the Holy Cross, in September."[14] Cattaneo has called this strategy a "kind of theater," and it's one where the audience is a participant.[15] And yet, some of these texts are downright defensive. Fra Mauro wrote about his placement of the source of the Nile River, "I think that many will be amazed that here I put the source of the Nile. But certainly, if they approach the question rationally and undertake the same investigations that I have—and with the diligence that I cannot here describe—they will see that here I am undertaking to demonstrate this thanks to the very clear evidence that I have had."[16]

Or, for all we know, Fra Mauro was a chatty person, and it was natural for him to talk and talk and talk about what he was doing and why.

In any case, lucky for us that he made this textual decision. The map would not be as spectacular, or as influential, without those writings. Directly on the map, the cartographer is taking to task previous geographers and philosophers and questioning bits of medieval thought about monsters and myths, all the while seducing readers with words to reconsider previous approaches and ideas.

THE LANGUAGE OF FRA MAURO'S MAP

Fra Mauro felt deeply that his map was intended for everyone, not just intellectuals and scholars. His intention is clear in an inscription on the map: "In this work, I have of necessity decided to use the modern names of common speech because, to tell the truth, if I had done otherwise only a few learned men would have understood me."[17] That particular inscription sets the tone for all the texts on the map and it invites everyone to take a look, no matter their education or intellectual ability. But even if Fra Mauro envisioned a wide audience, his final result was a map that few could interpret because the text is written in Veneziano, his native language. Trying to read it now requires fluency both in the Veneziano spoken today and medieval Veneziano. All languages change over time; they discard words, invent new ones, and change pronunciations, and Veneziano is no different. Today the choice of Veneziano might seem especially odd given the current restricted distribution of this language to the Veneto province of Italy, and as a second language to Italian at that. Although the word *dialect* is often used to categorize Veneziano and other regional languages of present-day Italy, it is a misnomer; they are separate languages with separate histories based on distinct regions.[18] Venice became part of Napoleon Bonaparte's Kingdom of Italy when it was passed from Austria to France in 1805, and then in 1866 it joined the real Kingdom of Italy. That means Veneziano was the dominant language of Venice and the area sounding the city, the Veneto, until a mere 150 years ago. In the history of Italy as a peninsula in the Mediterranean, that's not very long. As the introduction to Lodovico Pizzati's Venetian-English dictionary points out, millions of people still speak Veneziano every day. "From Verona, Lake Guarda to the rocky shores of Trieste and the Venetian Lagoon up to the peaks of the Dolomites, Venetian is the mother tongue."[19] This language was also carried to the New World by immigrants; many people in Brazil, Argentina, Australia, Canada, Mexico, and the United States still speak Veneziano. Pizzatti also points out that this language is the one spoken at home, with family, and only with those who already speak it, which cements a "confidential relationship." During the Middle Ages, Veneziano was also a widespread and popular language that had been transported across the world on trade ships, and by the Venetian navy, which was, at one time, the most powerful sailing force in the world. Because Venetians were the world leaders in international trade, Veneziano was once a common language in the West and the Middle East as the language used by sellers and buyers.[20]

Veneziano is also part of Venetian identity and the identity of the entire province that includes Venice, the Veneto, but it is an endangered language.[21] As a tourist in Venice, you hear standard Italian all over the parts of the city crawling with tourists. But there is an undercurrent of another language that a non-Italian might not even notice. For example, a quick trip on the public gondolas, called *traghetti*, that cross the Grand Canal for a small fee, is a great introduction to Veneziano because when speaking to each other, the gondoliers only speak in Veneziano and they are incomprehensible even to many who speak Italian. If you listen closely to locals on the street, in bars, or overhear street conversation, Veneziano is spoken all over the city; it still graces street signs and maps. The words are elided together to form new words and other words are cut off at either end so that to the untrained ear they sound like syllables or mouthed expressions rather than full words. Veneziano also has a musical cadence as the words slide into each other and the voice rises and falls, making individual words difficult to parse for nonspeakers. Veneziano is also known for its lavish use of *X*s in place of *Z*s but knowing this helps to spot Venetian words. Also, traditional Veneziano has no or few double letters, which serve as emphasis in standard Italian.[22]

Nonetheless, the choice of Veneziano for this map still seems odd given that scholars back then usually wrote in Latin, the accepted language of intellectuals, the educated, and the rich. Latin was also the language of the Catholic Church. But the choice of language may not have been an option for Fra Mauro. The Camaldolese order in the Venetian Lagoon was devoted to translating works into Veneziano so that everyday people could read manuscripts; this move was aimed at serving the populace, not just intellectuals and aristocrats. Therefore, it was probably not a huge decision on Fra Mauro's part, and for all we know it might have been required by the monastery. The original map was made for the Seignory, which means the government of the Republic of Venice, and so there, too, it would make sense that this map, as a product of Venice, would be in its native language to reinforce the might and sovereignty of the republic. Indeed, Venice has a long history of artistic works that acted as propaganda for the state, reassuring the populace of the power of their republic and unifying Venetians into an economic collective based on international trade.[23]

SPARING WITH, OR IGNORING, THE GIANTS

Since maps are representations of a cartographer's or geographer's world-view, they all come with a perspective, a viewpoint, a stance. Sometimes that perspective is purposeful in that the document is supposed to change minds; other times the purpose is to reinforce and reassure certain views. A Western medieval perspective was also informed by both religious beliefs and a need to project economic or political power. In other words, the history of world maps is a history of society as well as geography. As I outlined in chapter 2, by Fra Mauro's time, there was a long line of world maps from various places and the Venetian monk certainly knew much of that history. And so, Fra Mauro had to make some choices, choices about who to copy and who to challenge. Of course, he could have done this in silence. He could have just drawn the map, done the geography as he saw fit, and not explained a thing. But instead, he justified his acceptance and rejection of those who went before right there on the map, and some of those he was disagreeing with were traditional greats, which made many of his decisions geographical blasphemy.

Since Fra Mauro was a man of the West, he would have been aware of classical philosophers such as Aristotle and Socrates, among others, and his worldview was certainly influenced by their writings, one way or another. Ptolemy was a special point of contention for the friar. Claudius Ptolemy was a Greek mathematician, astronomer, and geographer (see chapter 2). Ptolemy's most significant contribution was to invent ways to alter systematically the coordinates of latitude and longitude using math and geometry so that the surface of a ball (the earth) could be laid flat without much distortion.[24] His second projection, the one that looks like a cape draped over someone's broad shoulders, and the one that became famous and was the standard for centuries, combined curved lines of longitude with curved lines of latitude. His projection was revered in Fra Mauro's day and in many cartographic centers, such as Florence, where aristocrats and libraries had copies of his eight-volume *Geography*, Ptolemy's view was sacred. But in Venice, the reception was not that reverent. Venice was a center for cartography in the 1400s and mapmakers and scholars were a bit more critical, and his ideas were apparently a problem for Fra Mauro.[25] The monk decided not to follow any of Ptolemy's projections, and he seemed to have some outright contempt for Ptolemy. Fra Mauro wrote on the map, "I do not think that I am being unfaithful to Ptolemy if I do not follow his Cosmography, because if I had wanted to observe his meridians, parallels, and degrees, I would have had to omit many provinces within the known part of the known part of the world that Ptolemy does not give: everywhere in his account, but especially to the north and south, he gives areas as *terra incognita*, because in his day they were not known."[26] In essence, Fra Mauro was saying that Ptolemy was missing giant swaths of the northern and southern parts of the world (let alone the whole other half of the earth, of which Fra Mauro was also ignorant) and therefore his projections and his ideas were antiquated and of no use. This statement was a slap in the face to Ptolemy and other cartographers who held him up as the standard for medieval geography. But no matter, because Fra Mauro was irked by Ptolemy's fame and thought his reputation was undeserved and that he could do better. For example, in one text on the map of Asia the monk claimed he was not naming all the Chinese provinces because the names had changed, but then snarkily adds, "But one should also note that I give other provinces that Ptolemy does not mention."[27] Mauro also does not hesitate to criticize Ptolemy for not knowing enough about Scandinavia, claimed he had no idea there was a Baltic Sea, and that Ptolemy was clueless about northern China.[28] Mauro also disses Ptolemy for stating in his book that he can only defend the names of places that had been visited often. Following that line of thought, Fra Mauro writes on his map, "So I say that in my own day I have been careful to verify the texts by practical experience, investigating for many years and frequenting personas worthy of faith, who have seen with their own eyes what I faithfully report here."[29] And yet, Fra Mauro was happy to use many toponyms and their geographical coordinates provided by Ptolemy. Of course, those names were initially written in Greek, and Fra Mauro did not know Greek, so he used the Latin equivalents, which are from a later translation of Ptolemy's *Geography*.

Fra Mauro also balked at Ptolemy's description of Africa when he chose to listen to living Ethiopians rather than the Greek geographer. Ethiopia was considered a Christian country, the largest one outside of Europe, and so it was normal that Fra Mauro had been exposed to Ethiopian views. There had been an Ethiopian "embassy," meaning a diplomatic mission to Venice, in 1402 and some Ethiopian Christian pilgrims had passed through Venice during that time. Another possible point of contact was the Council of Florence, which began in Basel, Switzerland, and was set in various European cities between 1431 and 1449. Its purpose was to join Eastern and Western Christians into one flock. In 1438 the council met in Florence, and in 1441 four Coptic Ethiopian clerics attended the conference upon invitation. They might have also traveled to Venice on their way to or from Ethiopia. Fra Mauro writes all over the continent of Africa that he has spoken directly with those who lived in Ethiopia, so historians assume it was this delegation of clerics. Or perhaps he talked with previous Ethiopian visitors to Venice. On the map, he repeatedly claims he received geographic information and drawings from Ethiopians in person. Since Ethiopia was the common name for all of Africa below Egypt at that time, Fra Mauro's information refers to many areas in Africa outside of present-day boundaries of the modern country of Ethiopia.

All these verbalized criticisms mean that Fra Mauro knew the powerful influence that the rediscovered and translated *Geography* had on cartography in the Middle Ages, and he wanted to come out from under that influence because he thought Ptolemy was wrong on so many levels. Also, his notes were the best way to announce that his world map was up-to-date and more accurate than anything from classical times, as it should be. He wrote on the map, "So if someone contests the work because I have not followed Claudius Ptolemy either in the form of the world or the measurement of latitude and longitude, I do not want to defend this map in any other way than that in which Ptolemy defends himself when, in the first chapter of the second book [of his *Geography*], he says that one can only speak correctly of regions that are vested continually; of those which are less frequented no-one should think himself capable of speaking with equal accuracy. . . . So I say that in my own day I have been careful to verify the texts by practical experience, investigating for many years and frequenting persons worthy of faith, who have seen with their own eyes what I faithfully report above." Fra Mauro's feelings about Ptolemy were quite modern, just like the bickering that goes on among academics in journals and at conferences. Still, he was making a bold choice and one that can be seen as brave.

Fra Mauro's battles with various earlier scholars are also evident in the texts that accompany his approach to the cosmos on the outside of the map in the four corners of the secondary frame. Surprisingly, his descriptions of the cosmos follow those of Ptolemy, which makes them clash with what he did on the map proper. It's as if Fra Mauro disconnected the geography of the world from the heavens and the universe and followed previous traditions, leaving theories about the cosmos to past philosophers and scholars of astronomy while moving the geography forward into the enlightened age on the map proper. Falchetta states, "There [on the map], Fra Mauro follows what one might describe as an open, composite method, which is rationalistic and 'scientific' *avant la letter* [before that term was coined] in that it places great importance on the information gleaned from eyewitness accounts of those who have traveled to the four corners of the globe. Furthermore, no theological consideration interferes with the geographical theory."[30] More pointedly, Falchetta says, "In short, one might say, that Fra Mauro is as 'experimental' as a geographer and cartographer, as he is a traditionalist as a cosmographer."[31] This is no surprise, especially for a monk. For Fra Mauro, as well as many other scholars of the time, there was an uncomfortable dance between understanding the natural word on its own terms, as nature, while respecting religious beliefs about those same pieces of nature. There was no theory of evolution yet, but this was the point when Western thinking had begun to move quietly away from faith-based approaches for explanations of natural phenomena. And on this map, we see someone who rose to that challenge by physically separating the natural world, the Earth, that could be seen and touched from what was beyond in the heavens, which could only be seen by looking up at the stars and planets with the aid of a telescope and imaging their structure and function.

In these cosmological notes, Fra Mauro addresses any number of issues related to the planets and the heavens. These explanations include the wonder of recurring rising tides, how land rose out of the water so that it could be inhabited, which climate zones are habitable, the size of planets, the distance of the planets, and how the continents are divided. In these notes, beyond Ptolemy, Fra Mauro cites any number of sources, including Aristotle, St. Augustine, Euclid, Avicenna, and Thomas Aquinas, among many others. For example, his drawing of the universe in the upper left corner as a set of concentric rings that illustrate the various layers of heaven and the planets with the earth as the central ball, is a design based on Ptolemy. In the same way, he writes near the diagram of the elements on the upper right about the moon's effect on the tides, which also comes from Ptolemy. Fra Mauro doesn't know exactly how that tidal action happens, but he emphasizes the effect of the wind on tides. In that corner, he also writes about the common understanding that water should be covering the whole Earth because land is heavy and should sink, and yet land appears anyway. Fra Mauro then turns to the Bible, the part in Genesis where the waters gather and the land appears, to explain how this happened—it was divine innervation that caused the land to rise within the vast seas. In this corner of the cosmos, according to Fra Mauro, it's the grace of God that brought forth the land, which in turn allowed humans to live. Here in these cosmographic bits, he sometimes turns to religion when there is no better explanation. The illustration on the bottom right depicts the earth as a globe hanging in space marked with "zones of habitation," which look like lines of latitude or climes. This Earth is half water, which agrees with the common idea that the known land took up one side of the earth while there was only water on the other side. The accompanying two texts are about sizing the earth and the universe. In these notations, Fra Mauro gives a history of ideas on these issues and contradicts both Aristotle and Euclid and maintains that all the climate zones, north and south of the equator, are habitable.

All the texts outside the world map seem to be academic discussions, much like the introduction to one of today's academic journal articles where a scholar must provide the historical context of whatever research they are presenting. Those introductions typically roll critically through the history of the subject, citing what is right or wrong about previous theories and context, and in doing so they set up new a hypothesis for testing. That test, the research at hand, is the main point of the article, but the basis for the research is previous research and previous theoretical approaches. In other words, these journal articles present historical context first and then walk the reader through the authors' new approaches, new ideas, and new data, which will then alter the conclusion of past work. One might think of the outer frame of Fra Mauro's map as the same kind of historical construct upon which his revolutionary map was based. In that approach, the outer parts of the mappamundi, as in the introduction to an academic paper, demonstrate to the viewer that Fra Mauro knew his stuff. Falchetta states, "One might almost suggest that these academic comments were placed around the world map to defend it from possible criticism. It is as if the author wished to demonstrate his command of the cosmological and geographical knowledge of the day so that any differences in his depiction of the form or extension of the *oecumene* [the earth] could not be taken as due to neglect—or even worse, ignorance—of the opinions of the learned."[32]

Fra Mauro's depiction of paradise on the lower left outside the map is different from the other three images because it is solely based on religious tradition, and because of that, it most dramatically contrasts with the map itself. For example, he explains next to his paradise, "And the Lord God planted a garden eastward in Eden: and there he put the man whom he had formed."[33] At the same time, Fra Mauro hedges his bets by quoting St. Thomas Aquinas, who had suggested that the Garden of Eden might be an allegory and not a true story.

These texts outside the map in the four corners appear to be a historically accurate recapitulation of the then-standard ideas of cosmological phenomena. In that sense, this part of the work is very distinct in its resource base and expression than the map proper. In a way, that juxtaposition between the cosmology and the geography is a perfect statement for a map that was echoing the transition from faith-based cartography to science-based cartography. The fact that this map was made by a monk also underscores his foot in both camps. Fra Mauro was obviously following the tenants of the Catholic Church, with all its stories about how God created this or that, but he was also a keen analyst of geographical information that was based in observable reality. His preference for eyewitnesses is a testament to the latter side of his nature.

In some instances, however, Fra Mauro backed away from possible confrontation with the past by not naming his sources at all. At first glance, one might think he was hiding from his decisions in these cases, a move exactly the opposite of citing various scholars and then refuting them. For example, the map has many toponymy and geographic features that Fra Mauro presumably learned from reading or hearing about Islamic maps, and yet he does not name any of those cartographers. It is possible that he did not know their names or those maps, and it is only through the hindsight of modern communications and access to knowledge that one assumes he must have been schooled in Islamic cartography. After all, he oriented this map to the south as had many Arabic cartographers before him, although we don't know if he was following those geographers or the few Western ones (including Andrea Bianco) who had previously made south-oriented world maps. But his use of Arabic place names is clear evidence that Fra Mauro knew something about the contributions of Islamic cartographers. For example, he calls the Maldives the islands Mahal and Duiamoal, which are Arabic names. And he echoes Islamic ideas when he puts the made-up canal between the tip of Africa and the rest of Africa. Also, his east coast of Africa is based on what Arabic navigators had found. At one point on the map, Fra Mauro also waxes lyrical about the dangerous whirlpool off the coast of Burma, and that must have come from Islamic sailors.[34] Since none of the Islamic maps had been translated and there is no evidence of them in the West, it's hard to know where Fra Mauro might have gotten this information if not from Arabic sailors and traders who passed through Venice while he was alive. He never specifically names any sailors or traders, but he repeatedly uses phrases like *those who have sailed here*. Or perhaps he viewed those resources so much part of navigational history, so much part of cartography—recall that the major Islamic world maps appeared between the ninth and early thirteenth centuries—that he felt no need to cite specific individuals or maps.

The not naming people, and not giving credit, was not limited to scholars and navigators from other parts of the world. Oddly, two of the most blatant omissions were well-known Venetians—Marco Polo and Niccolò de Conte—and he does not name the great Italian missionary and explorer Odoric of Pordenone, who came from an area just north of Venice.[35] Information from these three explorers is all over this map. For example, the toponyms for China and the route that the Polos took getting there and back are by and large directly taken from one of the Latin translations of Marco Polo's book. Fra Mauro even directly copies passages from Polo's *Il Milione* onto scrolls and inscriptions but does not give him any credit at all.[36] In the same way, Fra Mauro also uses the words of Odoric of Pordenone, who was traveling about the same time as Polo, and here, too, he never cites Odoric. The same with Niccolò de Conte, who supplied much of the information about Southeast Asia and the islands of the Indian Ocean. This lack of citation also sometimes makes him stumble cartographically. For example, for the summer city of the great Mongol emperor Kublai Khan, Fra Mauro uses the name Xanadu (a slight change from what Mongols called the city—Shangdue), which was the name that Polo used to denote that city, and Fra Mauro marks it with a hexagonal building and a pile of treasure. And he writes on the map, "To the admirable temple which is in this gulf come most of these oriental peoples to make their vows and offerings. This is why it is said that here there is such an accumulation of treasure that one cannot even estimate it."[37] But then Fra Mauro places another city name close by, Sandu, which comes from Odoric, but this is probably the same city. So, he was following both explorers and getting a bit mixed up in the process. But still, the absence of credit seems odd. Piero Falchetta suggests that Fra Mauro doesn't bother to cite Marco Polo, Niccolò de Conte, and Odoric of Pordenone because they were from "ancient times" and what they wrote or said was common knowledge, accepted by everyone, and therefore not in need of a citation.

In another odd move, Fra Mauro's most treasured sources, those many eyewitnesses "who saw with their own eyes," are also never mentioned by name. He does this with reports from sailors, explorers, and travelers. They are only characterized by the sweeping phrase *many have told me*. In that case, he may have thought this was recent news and so he lets the viewer decide if they want to accept or reject those descriptions. But by failing to name these faceless informants, the map takes on an aura of immediacy, and perhaps that's what Fra Mauro was aiming for. Accept it or reject it as you see fit, he seems to be saying. As to who these people might be, it's easy to imagine the various visitors to the monastery who came for a chat with the priest or to picture Fra Mauro walking through Venice and perhaps standing at the market and talking to a stranger who had just arrived. They might have discussed where that stranger came from and how they got there and, during that conversation, Fra Mauro may have failed to note their names because he was concentrating solely on the information they were imparting.

THE LAND

Space was an issue for Fra Mauro. Yes, his map was gigantic at seven feet in diameter, but it was still limited. To accommodate that space, he fiddled with the standard shape and size of the continents, but for good reason. As he pointed out in a legend, "Let people not be surprised that in Europe I have shown cities so small and in Asia so big. Where I have had space I have made the places big; where I have been short of space, I have made them small. Let those who see them bear with me if they do not find them totally satisfactory and fully as they would want."[38] But he also found the very issue of the continents boring. After all, the vision of a three-continent world had been around for a very long time. He wrote on the map, "Regarding the divisions of the world—that is, of Asia from Africa and of Europe from Asia—cosmographers and historiographers give various opinions. Of these, one could discuss at length, but because it is boring to dwell on this controversy, I will make a brief note with their opinions and leave the prudent to decide which one they should hold as best. . . . However, I advise those who are looking at this work not to worry themselves too much about discussing this division, given that it is not that important. Let them opt for that which seems to them most reasonable and probable, both to the eye and to the intellect. None the less, I remind them that it is a praiseworthy thing to follow the authority of the most veracious."[39] Fra Mauro was also trying to fit all the land into his planisphere while also making room for all that water. And so, he changed the size of landmasses, even shrinking deserts because they weren't habitable; in his human-centered mind-set about the map, he didn't care about empty deserts. Fra Mauro was mostly interested in the world as it was inhabited by people. This wrinkle in his perspective was probably a result of his religious training that puts humans as the descendants of Adam and Eve in charge of all other creatures as the "owners" of the earth.

Fra Mauro was also worried about the ability of his viewers to dissect and comprehend the topography of his map. He wrote, "Note that throughout this work there are some green marks or some rows of small trees which serve to indicate the divisions and borders between provinces. But those who want to understand well must have seen with their own eyes or else have read well. And they must have a good sense of proportions and the ability to interpret the drawing; otherwise, they will draw little fruit from these signs and will not be able to form a good idea of what they see depicted above."[40] Since Fra Mauro had not traveled much outside of Venice, he must have thought of himself in the latter category of viewer.

AFRICA AND ROUNDING
THE CAPE OF GOOD HOPE

The Africa on Fra Mauro's medieval map is not the Africa we know. It takes up a large part of the top right of the map (or the southwest when oriented north) and it's an oddly shaped landmass that curves to the right at the bottom seemingly to fit into the circular frame. The Great Rift Valley, the birthplace of humankind and Africa's grandest geological feature that begins in Lebanon and cuts through the full length of East Africa from Ethiopia to Mozambique, is simply not there, presumably because no Europeans or Arabs had ever seen it. The Red Sea is hefty and accurately drawn, as is the Gulf of Aden; there is also the shadow of the Horn of Africa as we know it today. On the western side of Africa, this map has a giant gulf about one third down from the top of the continent and another smaller gulf two thirds further on. And there is a very strange watercourse, a river, or maybe a channel, because unlike a river it barely meanders, running northeast (or horizontally if you reorient the map to make Africa straight), slicing off the tip of southern Africa. We see here an Africa coming into shape, but it isn't quite accurate yet. Still, Fra Mauro had made major and revolutionary geographic decisions about Africa that require explanation.

If someone knows about Fra Mauro's map, they usually know that its major contribution to world maps was the first clear purposeful depiction of the southern tip of the African continent showing the possibility of sailing from the Atlantic Ocean into the Indian Ocean. Other world maps before this one usually sported a truncated Africa, as if half the continent had been sliced away. Some others either push southern Africa right into the frame, which represented the end of the world or, as in Ptolemy's maps, curve it sharply right and attached the lower tip of Africa to the Indian Peninsula. But there were also ancient stories about the possibility of rounding the southern tip of Africa from either direction. For example, Eudoxus was a Greek explorer who at least twice in the second century BCE started from Greek-occupied Egypt and sailed down the Arabian Sea, out across the Gulf of Aden, and then using the Indian monsoon winds, sailed further into the Indian Ocean. Eudoxus also did some trading in the East and brought back to Egypt gems and other goods, so presumably, these were trading expeditions. But during the second voyage on this route, so the story goes, Eudoxus and his crew were blown off course by the terrible winds in the area and their boat was sent around the Horn of Africa and further down the east coast of Africa instead of out to sea. Somewhere along that coast, Eudoxus and company found a shipwreck that was said to be left by a crew that originated in Spain and sailed down the west side of Africa, rounded the tip of the continent, and sailed up the other side of Africa until it shipwrecked. Assuming he could make the same trip but in the opposite direction, Eudoxus supposedly sailed down the east coast of Africa as well and tried to round the tip going south and then west, but no one knows if he ever made it.[41] There were also anecdotal reports of Arab sailors going back and forth below the tip of Africa, but again, there is no documentation of these voyages. For centuries, Arab sailors ruled the Indian Ocean and for centuries they had been peacefully trading in those waters, known to Arabs as the Sea of Zani, going as far as the island of Mauritius and just east of Madagascar, both close to the African coast. But until Fra Mauro, there was no documentation of any of this and everything was speculation.

Fra Mauro was also not alone in this possibility on maps. Previously, world maps by Pietro Vesconte (1325), the *Medici Atlas* (1351), and the de Virga world map (1415) had water around the southern part of Africa, but it's hard to tell if these represent circumnavigation or simply the old idea that the world was surrounded by a ring of water. There was also the drawing of Africa by collaborator Andrea Bianco (1436) with the southern tip of Africa broadly curved in an easterly direction, almost to India, and a scrim of water beneath that also might have been representative of a border or maybe real navigable water. Instead of citing these resources (and perhaps he was only familiar with Bianco's map), Fra Mauro moves to clarify that he is stating that Africa is circumnavigable. In doing so, he leans on his usual fallback—favoritism for eyewitness reports over historical cosmologies or historical projections.

The possibility of rounding Africa and finding open water on the other side was of prime importance because that route was the Holy Grail for international European trade. That trade was ramping up in the late Middle Ages as shipbuilders began to design and make boats that could carry huge loads, navigational charts had improved, and capitalism had blossomed in Europe, especially in Venice.[42] The ability to leave the Mediterranean, sail west out into the Atlantic, south down the west coast of Africa, and then slip around the southern tip and into open waters on the eastern side of Africa was a coveted, and controversial, idea during Fra Mauro's time. And yet, no European had gone further south down the western side of Africa than Cape Bojador on the coast of Western Sahara until the Portuguese, on the instigation of Henry the Navigator, moved farther south to Râs Nouâhiboud (Cabo Blanco) in present-day Mauritania in 1441.[43] When Fra Mauro was working on his map, explorations down the coast had not yet even covered the top western half of Africa, the rounded bump. For a very long time, there seemed to be some kind of psychological barrier stopping expeditions from going further south to the tip. That hesitance was probably fueled by previous maps that had truncated Africa and inhabited it with monsters, combined with those that decried that the lower part of Africa as terra incognita and an empty wasteland. In addition, the first voyages down the western side of Africa turned out to be unprofitable; the idea was to find gold and gems, maybe even spices, but there was none of that in northern Africa, so why go further? Portuguese navigators did do some trade in ivory and they enslaved people and transported them from the continent, but the better financial gain would be making it to the East and sailing back. It would take until 1488, long after Fra Mauro was dead, for the Portuguese navigator Bartolomeu Dias to go all the way down to the tip of Africa and swing left and around the tip.

But why was Fra Mauro so confident that Africa was its own continent and not connected to Asia? And why did he believe so deeply that Africa that could be circumnavigated? The answers to those questions appear right on the map where he relies heavily on Portuguese explorers and their maps, which Fra Mauro alleges to have seen in person. He wrote,

"Many opinions and many texts claim that in the southern regions [of Africa] the water does not surround this whole inhabited and temperate area. But I have heard many opinions to the contrary, above all from those who were sent by His Majesty, King of Portugal, in caravels so that they might explore and see with their own eyes. These men say that they sailed around the coast from the southwest some 2,000 miles beyond the Strait of Gibraltar. And following that route, they then decided to sail south-southeast until they came into a line with Tunis and almost as far as Alexandria; and at each place, they found good shores, with deep water and good navigation condition without any hindrance. These men have drawn new navigation charts and have given names to rivers, gulfs, capes, and ports, of which I have had a copy. So if one wants to disagree with these men, who have seen with their own eyes, then there is all the more reason to descent from and not believe those who have left writing on things they did not see with their own eyes but only believed to be the case."[44]

He also repeats the stories of another sailor "worthy of trust" who had been on an Indian ship going southwest that was caught up in a storm for forty days, and some astrologers (meaning astronomers) who were charting their course and reported that they had gone about two thousand miles when the storm had pushed the ship another two thousand miles. Then he retells the story of Eudoxus, which he got from reading Pomponius Mela, that a sailor had left Alexandria and sailed out of the Arabian Gulf and went around the tip and ended up in Gibraltar. Fra Mauro also recounts the story of a ship or junk that in 1420 began in India and sailed to the Island of Women (which was next to the Island of Men; see chapter 6) off the Cape of Diab. According to the story, this boat sailed forty days in a southwesterly direction and found nothing but the blowing wind and lots of water. The ship apparently rounded the cape and went two thousand miles up the west coast of Africa and then went back to the tip. He also cites Pliny, suggesting he had heard of two ships fully loaded down with spices that had sailed down the Arabian Sea and around Africa to Spain where they unloaded the cargo at Gibraltar, a story repeated by one Fazio degli Uberti, a poet of fourteenth-century Florence whom Mauro felt was a trusted resource.[45] He might have also known that in the fifth century BCE, Herodotus wrote that Africa was circumnavigable. With these stories, he takes the stand that Africa can be circumnavigated.

Then Fra Mauro dropped the other shoe, one that few had considered: "One can therefore claim without any doubt that this southern and south-western part is navigable and that the Sea of India is an ocean and not an inland sea."[46] That simple sentence changed the Western view of the world as it "revealed" another ocean, and the idea was revolutionary. In another inscription on the map of Africa, Fra Mauro tried to lay out support for this notion. "Some authors write that the Sea of India is enclosed like a pond and does not communicate with the ocean. However, Solinus claims that it is itself part of the ocean and that it is navigable in the southern and southwestern parts. And I myself say that some ships have sailed it along that route."[47] What was so life-changing about this section and the drawing of the map for both Africa and the Indian Ocean was that contrary to Ptolemy's belief that Africa was attached to the Indian Peninsula, there was a water route from Europe to the East, which completely opened up that part of the world for exploration by water . . . and the water at hand was an ocean.

But the Fra Mauro map of Africa also has glaring faults. By his time, the eastern side of Africa had been well documented on maps by Arabic cartographers and navigators, and the Coptic clerics from Ethiopia had filled in many spaces on Fra Mauro's Africa, making that part of the continent reasonably accurate. But his western side of Africa was scarred by a gigantic, oddly shaped, and indented Gulf of Guinea, which he called Sinus Ethyopicus. The vastness of that water space on the western side of Fra Mauro's Africa carves deeply into the northern half of the continent and we know it's not accurate at all. The top left of Africa is drawn as a broad curve that cinches the land halfway down and forms that bulge at the western top of the continent, and it cuts east into the African interior. Some historians have suggested that Fra Mauro's view was influenced by descriptions and drawings of the various powerful rivers that dump into the Gulf of Guinea, which might suggest to some sailors that there was deeper water behind their grand deltas. He also might have been influenced by tales of Emperor King Prester John and his imaginary kingdom (see chapter 6).[48] The idea was that this giant gulf somehow led all the way to central Africa and so might have been a passage that reached Egypt and the Christians under Prester John's rule. To be able to navigate that fanciful passage would mean avoiding the Muslims in North Africa, who were supposedly threatening the Christian faithful of Prester John. Mauro's Gulf of Guinea is also dotted with islands, and he adds, "on some of them live Christians," supporting the notion of scattered and venerable Christians in Africa.[49] Or maybe he was just taking a page from the de Virga world map, which shows a Gulf of Guinea so large that it topples North Africa downward to the west. Or he was following the large bays on the Vesconte and Bianco maps that appear on the west coast of Africa.

The other geographic conundrum on this African map is the shape and separation of the tip. Fra Mauro drew it as a large triangular piece of land at the southernmost point that was disconnected from the mainland by a blue stream of water. Diab was the Arabic name for "this piece of land" in documents of the age, but it's not clear exactly what this large chunk of land and that line of water might be. Al-Bīrūnī, the Arabic humanist, mathematician, and geographer, had suggested in the eleventh century that there was such a channel on that part of the African continent. He, like others, believed that the African continent bent east to connect with Asia. A channel, al-Bīrūnī felt, allowed Arabic sailors to pass from the Indian to the Atlantic Ocean, like a human-made canal with locks. And some Islamic maps did indeed show the southern part of Africa curving to Asia with a channel that would connect the Indian "Sea" to the Atlantic Ocean. In other words, al-Bīrūnī and some Islamic cartographers were suggesting that Africa was "circumnavigable" by way of this natural channel.[50] And yet, that so-called channel might have been simply the tip of Africa and the so-called land on the other side of the channel assumed, but never seen. It would be expected that Fra Mauro knew of al-Bīrūnī's writings, and he may have been influenced by the explanation. Fra Mauro writes about that line of water that disconnects it from the main body of Africa, "Note that this Cape of Diab is separated from Abassia [the rest of Africa] by a channel that is lined on both sides by high mountains and trees that are so tall and thick they make the channel dark. The water within it forms a whirlpool so dangerous that any ship that encounters it is in peril."[51]

Fra Mauro seemed to think this extra bit of land at the base of Africa was the tip of Africa, but it's hard to tell because of that line of bright blue water that separates it from the mainland. One possibility was that if you took that chunk of Diab and swing it up the east coast of Africa, it could be the island of Madagascar, which lies on the southeastern corner of Africa across from Mozambique. When the Venetian writer Giovanni Battista Ramusio was compiling his three collections of the travel writings of others, called *Navigazione*, between 1550 and 1559, he

considered Diab to be Madagascar.[52] But Placido Zurla, a fellow Camaldolese monk who did the first explanation and study of Fra Mauro's
map in 1806 (see also chapter 7), rejected any thought of this "island" as
Madagascar.[53] He maintained that it was indeed the tip of Africa. And
yet modern historian Piero Falchetta agrees with the idea that Diab
might be Madagascar because Fra Mauro describes those well-known
whirlpools near Diab that could swallow ships, and those whirlpools are
actually near Madagascar. Mauro also comments on a cave that is found
on the east side of Madagascar.[54] And Falchetta notes that the word
Diab comes from another Arabic word that was used commonly to refer
to all the land of southeastern Africa, and so Fra Mauro might have been
using it in that generic sense.[55]

No matter what Diab was in Fra Mauro's mind, all these thinkers,
navigators, and geographers were underscoring the idea that it was possible to sail along the southeastern side of Africa and then turn west-
ward (turning right, as it were) and continue sailing into to the Atlantic,
through a channel or around a tip. Arabs surely knew, although we don't
know exactly how, that Africa could be circumnavigated, and perhaps
they were simply inserting a channel to accommodate the idea that was
already in established Islamic maps that Africa and Asia were connected
by land, and so there must be a channel that allowed a boat to pass from
the Indian Sea to the Atlantic Ocean. On the other hand, they might
have been describing the Mozambique Channel, which separates Madagascar from Africa.

Fra Mauro's Africa is crowded with toponyms; that one continent sports the lion's share of the place names on the map for a single
continent—585 out of a total of 2,921—but much of the interior is blank
because Europeans, and other non-Africans, had not yet gone there. Fra
Mauro also chose to note ancient Egyptian and Roman settlements. He
received many of these place names from that delegation of the Coptic
Ethiopians (see also chapter 6) as they apparently met with Fra Mauro
in person at some point. On the map he writes: "Because to some it will
appear as a novelty that I should speak of these southern parts, which were

almost unknown to the Ancients, I will reply that this entire drawing, from Sayto upwards, I have had from those who were born there. These people are clerics who, with their own hands, drew for me these provinces and cities and rivers and mountains with their names; all these things I have not been able to put in due order for lack of space."[56]

Interestingly, Fra Mauro often uses Arabic names for places in Africa, but the Venetian cartographer Albertinus de Virga had already used Arab toponyms on his world map. And Fra Mauro was surely versed in these names as known to Arab sailors who must have visited Venice at some point.[57] The area around the course of the Nile River is especially detailed, which was of great import for that time, and a controversial subject, for cartographers. On one inscription near the Nile, he explains that what others had called the source of the Nile was just a secondary branch of the great river, "because within it one finds animals similar to those in the Nile."[58] Said like a modern biologist.

THE EASTERN (INDIAN) OCEAN

The Indian Ocean and its islands and coastlines as drawn by Fra Mauro were important because this is the area of the spice trade from which Venetian merchants had made their wealth. These trade routes were not conducted along the Silk Road, which involved years of hard land travel, but instead over water among the islands of the Indian Ocean and at ports along the coastlines of India, Southeast Asia, and China. Trade here had been established by Arabic traders who then brought goods to port cities in Asia Minor on the Black or Caspian Seas where they could trade with Europeans. But that "sea" was an ocean, and it was virtually unknown to Europeans except for hearty exporters such as Marco Polo and Niccolò de Conte. Unlike Africa, this part of his geography is less revolutionary and less idiosyncratic; instead, it aligns with Ptolemy, which also engenders some mistakes, such as making India strangely shaped and Sri Lanka way too big.[59]

While mapping the area of the Indian Ocean, Fra Mauro continued
to rely on first-person information. Much of that information must have
come from Arab sailors and traders, presumably those who passed through
Venice and may have had direct conversations with him. Fra Mauro repeatedly underscores whatever he says about the area of the Indian Ocean
with the phrase *those who sail in these seas*, and those witnesses can only be
Arabic traders.[60] He was also schooled in Arabic geographers and their
thoughts, and he surely knew the travels of Ibn Battuta from the fourteenth
century, although he may have hesitated to give them credit on the map
because such knowledge might have been considered "outside the canon"
for a Catholic monk.[61] He must have also worked from a base of Arabic
navigational charts (portolans), which would have contrasted somewhat
with Ptolemy's view of the area. He strongly maintains that this ocean is
"internally navigable" and not the scary unknown swath of deep blue sea,
the *mar scuro*, represented on other Western maps. On this map, he also
constantly refers to various established trading routes across the Indian
Ocean and the products the various islands offer (see also chapter 6), which
underscores how he expected the map to affect the future of international
trade. Fra Mauro also points out that this area of endless water is full of
human life and activity. Historian Marianne O'Doherty writes, "On the
map the Indian Ocean world comes across as positively crowded; populous
and well-traversed, it is a cog in the economic machinery of the world."[62]
Fra Mauro further claimed there were over twelve thousand islands in the
Indian Sea and that the inhabited ones were fertile.[63] Even if this part of
the world was so vast, Fra Mauro ran out of space for everything: "In this
sea there are many islands that cannot be specially noted because of lack
of space."[64]

And yet, he goes on to describe many of the islands in detail, often
including the products grown or made on these islands, or the practices
of various peoples (see also chapter 6). A few islands merit special attention with several inscriptions and detailed drawings, such as Sri Lanka,
and Sumatra, which is now part of Indonesia. The name "Taprobana"
had been used by others before him to denote Sri Lanka and Sumatra,

so that's a bit of confusion on the map. He uses the word "Saylam" for Sri Lanka and "Taprobana" (as well as the name Siamotra) for Sumatra, which follows Marco Polo.[65] In fact, for the Indian Ocean and the surrounding area, Fra Mauro relies heavily on the descriptions and place names given by Polo and Niccolò de' Conte. De' Conte's travels were written down by Poggio Bracciolini in 1444 right before work on this map began, so the monk would surely have considered those words as eyewitness accounts. Sometimes he chose one Venetian explorer's version over the other, although we have no idea why.[66] He claimed that Sumatra was more than four thousand miles around. Fra Mauro's Java is oversized, as it would be since this island was so important in the spice trade. To the west of Java is a straight string of islands that reach down to the frame, and of course, there is no such straight line in the real world. Instead, it seems to be depicted as some island barrier, or the line might just represent one possible formation of the myriad islands that traders had visited while gathering spices. Fra Mauro also drew two small islands surrounded by flames in the Sea of Aden, which flows into the Arabian Sea, part of the present-day Middle East, but these waters are also part of the Indian Ocean. He notes that "those who sail in these seas say that these two mountains burn," which means they are volcanoes.[67]

Fra Mauro also gives warnings to those who might want to venture off the trading path. For example, he points to a whirlpool on the eastern side of the Bay of Bengal, which he heard was dangerous for sailors.[68] Following Arabic navigators, he claimed that there were shadows and darkness in the southern part of the Indian Ocean. "The ships which, sailing south, pass too close to the Lost Islands are carried by the currents into the Shadows, which—like the waters below them—are so dense that the ships cannot go forward, and they inevitably perish."[69] On the map he also has a string of small islands right at the edge of the frame that starts at the tip of Africa, follows the edge of the map under the Indian Ocean, and then swoops up to Hangzhou on the east coast of China. In an inscription, he says these islands have all kinds of different birds, but more importantly, that if one were to pass these "barrier" islands, the ship would be stuck in the Shadows forever. "And this is known from experience of those who did venture into them, and who perished."[70] Charmingly, Fra Mauro reiterates from Marco Polo that once you get to Cape Chomari in India, you lose sight of the North Star because you are too far south.[71]

The names of places inscribed on the map of the Indian Ocean are a mix of endemic languages written phonetically (for recent research on this issue, see chapter 8), mostly from Marco Polo but also from Niccolò de' Conte, and a sort of Latinized version that echoes Ptolemy. The larger parts—that is, island and water sections—are written in a Latinate while the local names are in attempts at local languages.[72] With this decision, Fra Mauro is giving Western viewers names they might try to pronounce, but he was also staying faithful to the idea that other places have other ways of naming their environment.

ASIA

This vast content begins in the west where the Eurasian landmass is traditionally divided into Europe and Asia by the Volga or Don Rivers and extends all the way east to the China Sea. On Fra Mauro's map, Asia encompasses what was known as Asia Minor and the countries that are now included in the area known as the Middle East. Asia also spreads on this map from the northern reaches of Russia that end at the Barents Sea down to the Indian subcontinent, Southeast Asia, and the Middle East, which terminates at the Arabian Sea in the south. For the geographic part of the map of Asia, Fra Mauro relies on Ptolemy. Also, much of that grand landmass was not known to Europeans and there are empty spaces here and there. As for toponyms and descriptions, Fra Mauro relies heavily, once again, on the descriptions by Polo and de' Conte.

Although there is much that is recognizable today on his map, Fra Mauro also had a hard time with various parts of Asia.[73] For example, he grappled with the rivers of Asia as they wiggled across the landscape and emptied into the Indian Ocean, as well as the China, Barents, Caspian, and Black Seas. Among the rivers he tried to trace are the Ganges, Indus, Yangzi, Tigris, and Euphrates, all major rivers, and he often got the courses of these rivers wrong. It is especially odd how off he was about the Indus and Ganges since there were well-known at the time.[74] Perhaps it was just too much. On the map he names thirteen rivers but ends with "and of many others that cannot be given special note because they are infinite in number."[75]

Starting from the west end of Fra Mauro's Asia, one might expect him to focus a bit on Constantinople, given that city's influence on the history of Venice. Byzantium ruled Venice for a long time and was a major trading partner even after Venice became an independent republic. But he only mentions the city in one inscription: "This most noble city of Constantinople was in ancient times called Byzantium, but later it was extended by Constantine, who transferred the Roman empire there."[76] Constantinople, and thus the Byzantine Empire, fell to the Turks in 1453 which put the Ottoman Empire ever closer to the possibility of invading Europe, and Venice. Fra Mauro was in the middle of working on the map when this happened and he could have taken that possibility into account, but he didn't. Or perhaps Fra Mauro was finished with that part of the world when Constantinople fell, and he so moved on. But still, it's an oddity with no explanation. He does show a lot of interest in the area northwest of Constantinople broaching the Adriatic. Today the countries in that area include Greece, Bulgaria, Albania, Serbia, and Croatia. They interested Fra Mauro because they were near home, and he took the opportunity to point out where the first Venetians came from. For example, in one inscription he writes about the Greek province Paphlagonia and that Trojans made their way to the Veneto, the province of Venice, and settled in the area, making them the first Venetians. Asia Minor also held significance because so much Venetian trading happened there, especially around the Black Sea, where Venetian traders had permanent trading posts, so that part of the world was reasonably well-known to Europeans. But all that detail became too much: "Note that in this Asia Major there are many kingdoms and provinces which I have not put because of lack of space. Thus, I have decided to omit many things and take out those that seemed to be best known. And I have not even made mention of many rivers, mountains and deserts in many parts. . . . Summarily, I have decided to say nothing about the novelties, customs, and standing of the various people, about the magnificent and powerful seignories, about the great diversity of animals, or about an infinite number of other things."[77] But he guards against some possible criticism for places well-known to others: "If to some it seems that I have not located Babylon well, because I have shown it on the Tigris and not the Euphrates, as the Authors write, may it please them to look first at the drawing and then ask those who have seen with their own eye; they will thus understand that I do not stray from the truth."[78]

In the present-day Middle East, Fra Mauro also omits places that are well-known. Near Israel an inscription reads, "Those who expert, let them complete this Idumea and Palestina and Galilea with what I do not put—that, the river Jordon, the Tiberian Sea, the Dead Sea and other places which it has not been possible to include."[79] This is not so much a slight as it is a nod to those places that were well-known by European traders and crusaders. That type of omission would continue in his depiction of Europe, where he seemed to feel that notes and explanations were simply not necessary because anyone who looked at this map would already know everything about these places.

Mapping China was an entirely different challenge. Several areas of China were made smaller, and they contain few details, although there are many illustrations of towns, castles, temples, towers, and walls, all taken from Polo and de' Conte. Keep in mind that there is not a single human figure on this map, only the buildings, roads, and features that people have constructed. His China bleeds into Russia at the northernmost (or southernmost if the map is oriented according to Fra Mauro) part of the Asian continent and it is full of lakes. He divides Russia into White, Black, and Red provinces based on their provenience near rivers with the same colorful names. But, in general, he seems ignorant about this most northern part of the globe, as he moved places that are inland to the coastline. He does acknowledge that the Barents Sea is navigable, but most of the time he seems lost. Piero Falchetta states, "In this part of his map, the drawing and overall 'vision' of the world cease to be inspired by objective geographical considerations. Fra Mauro's account of Northern Asia, therefore, is a mere fossil, inspired by nothing more than the habits of thought that had informed the cartography of the Middle Ages."[80] But rounding along the northern coast and down to the eastern shore of China Fra Mauro planted a very modern flag—he put Japan correctly on the eastern shore of China, calling it the Island of Cimpagu, which was the first time Japan appeared on a Western map.

EUROPE

Fra Mauro's Europe is small compared to other medieval maps, which should be compared to Africa and Asia. What he presents here is a modern proportion, and one that is geographically accurate. Fra Mauro knew that a small Europe would be hard for proud Westerners to swallow, but no matter. He took the same tack with the Mediterranean and its many islands, which were very well-known to Europeans, especially traders and sailors. "Note that in this sea there are numerous islands that the *auctores* speak of a lot. But here I have only given the main ones, the noble character of which I cannot describe because of lack of space. Those who are scholars may make good this omission."[81] He dismisses the land that borders the Mediterranean in the same way: "Though I have been most diligent in trying to put all the coastlines of this seas in accordance with the most accurate map that I could, those who are experts should not take it amiss if I am not always consistent. Because it is not possible to put everything accurately."[82] He even passes over Italy in the same glib way: "Here I do not say more of this most noble Italy because it is so famous and has been celebrated by many most gifted writers."[83]

In the few explanatory inscriptions on Europe, Fra Mauro notes that Sicily was joined to the peninsula of Italy but the "force of the sea" then divided it from the European continent.[84] He also says that the so-called columns of Hercules are simply the mountains on either side of the Strait of Gibraltar.[85] Going into the north of Europe he claims that Scotland is attached to England, but that there was once a river between the two and that there are some mountains that now separate them. This hedging is probably a leftover from knowing about Hadrian's Wall, a defense structure built by the Romans, but it wasn't built at the current Scottish border.[86] He also notes the Shetland Islands in the North Sea and calls them Solan,[87] and he points out the Faroe Islands, which are even farther north than Shetland.[88] To the west of the Faroe Islands, he also calls one large spot Ixilania, which is Iceland.[89] Fra Mauro was aware of much of the Far North from the travels of the Venetian Zen brothers in the late 1300s,[90] and also from the story of the great Venetian explorer Pietro Querini, who shipwrecked on the Lofoten Islands in 1431 and then traveled across Norway and Sweden.[91] Fra Mauro's Scandinavia hangs over much of Europe, and his Baltic Sea nicely separates Scandinavia from Russia and the northern provinces of Europe. Venetian sailors and traders knew these waters, so it's no surprise he had most of this area reasonably correct, although he shows Denmark as an island.

WHAT FRA MAURO WAS COMMUNICATING
ABOUT GEOGRAPHY

The overarching geographic effect of this map is that it pictorially, and verbally, expands the world of the Middle Ages. It's also clear that Fra Mauro concentrated on the places where humans had inhabited the earth. That focus shows up in many of the images on the map, and in the endless inscriptions that comment on the vagaries of human behavior (chapter 6). This focus is also underscored by the cosmological drawing on the bottom right outside the map that shows that half the globe is covered with water and therefore uninhabitable. His text by that globe points out that he felt more of the world was habitable than previously thought and that people could live in climates that seemed too hot or too cold to the medieval mind. In other words, a focus on human habitation is one of the most important orientations of this map. Fra Mauro was also interested in human movement. He was adamant that people could easily sail from one place to another, as in rounding the tip of Africa and entering the Eastern Ocean, where many goods and glories were waiting. With a nod to international trade as the commercial advantage to exploration, Fra Mauro also depicts eight different kinds of boats across the waters, some from non-Western ports and barrels and big boxes presumably full of trade goods floating in the seas. This is, after all, the map that launched the Age of Discovery, and it appears that Fra Mauro intended that end. He basically opened his arms and said, "Here it is, you can get to it all, and the habitable world is not as scary as others have told you."

However, the geography presented on this map shows a distorted half-world to the modern eye. We see only half the earth, no North and South America, Australia, New Zealand, Pacific Islands, not even Antarctica. Although people were living in all these places (except, presumably, Antarctica), no one in Europe, Africa, or Asia had any idea what was on the other half of the globe, and yet they suspected there might be something westerly into the Atlantic Ocean. But the geography of Fra Mauro's map is a record of what was known through exploration and trade by the mid-fifteenth century. This map also pushes that exploration further by proposing an Africa that could be circumnavigated which would connect the Atlantic and Indian Oceans and make the East much more accessible to European traders and vice versa. Of course, to our contemporary sensibilities, a trip on a sailing ship that takes months or years doesn't seem fast, but in those days, a water route was still faster than traveling east overland. Transporting goods by land across Asia and Southeast Asia to where they could be loaded on ships in the Mediterranean or the Black Sea was time-consuming and arduous and added to the cost of trading. It was more profitable to go by water in boats owned by merchants or governments than to build networks of suppliers and middlemen to meet ships at ports in the Levant and elsewhere. Land routes also meant that traders had to utilize ports in foreign cities for storage, so shipping directly from some eastern port and having the opportunity to sail those goods directly home was ideal. A water route to Asia was just plain better.[92] Fra Mauro's map showed visually that the dream of a water path to the east was possible and doable, and his geography made it so.

But Fra Mauro's map is not just a map, it's a discourse. By his use of local place names, Arabic navigational charts, the voices of Arabic sailors and traders, and the geography of Coptic Ethiopian clerics over Ptolemy, he was also taking a step toward a sort of medieval global inclusion. That inclusion is even more apparent where Fra Mauro describes the people and places seen by Marco Polo, Niccolò de' Conte, and other Westerners, and includes their observations on the map—including the native nomenclatures they provided. Those descriptions of people and their habits are the next layer of discovery on this map.

CHAPTER 6

Peoples, Goods, Myths, and Marvels:
Lessons from Fra Mauro

*Here you cross a number of fine streams, and see a country
covered with date palms, amongst which are found the
francoline partridge, birds of the parrot kid, and a variety
of others unknown to our climate. At length you reach the
border of the ocean, where, upon an island, at no great
distance from the shore, stands a city named Ormus, whose
port is frequented by traders from all parts of India, who
bring spices and drugs, precious stones, pearls, gold tissues,
elephant teeth, and various other articles of merchandise.*
—Marco Polo, *Il Milione* or *The Travels of Marco Polo*, 1300

*So I say that in my own day I have been careful to verify
the texts by practical experience, investigating for many
years and frequenting persons worthy of faith, who have
seen with their own eyes what I faithfully report here.*
—Fra Mauro, Mappamundi, 1459

All things should belong to one world.
—Fra Mauro, Mappamundi, 1459

Mappamundi have been called the encyclopedias of their age.
Beyond what they reveal about how various cartographers envisioned the known world, these world maps were also overlaid with
other, geographically extraneous, information that made them much
like modern knowledge aggregators.[1] The notations on the maps, the

illustrations beyond landmasses and watercourses, and the texts written on or outside the maps were reflective of accepted philosophical texts, the travels of explorers, and lots of handed-down myths and exaggerations. Religious orientation also had a fundamental influence on how these maps were drawn geographically as well as the cultural overlay. Religion determined their orientation, and what was included, omitted, or emphasized. For example, mappamundi were documents of persuasion, or reiteration, for Christianity in the West. Mappamundi were not just depictions of land and sea, they were also products of various cultural, ethnic, political, and religious biases.

Because of that complex history and purpose, today these very same historical documents are useful as we look back and try to comprehend human behavior, thought, and perspective over the centuries. What makes them even more special is that they are both pictorial and textual. That's why the inscriptions and illustrations on Fra Mauro's map that go beyond geographical features are so important. They are the heart and soul of this map and a history lesson about human knowledge and perspective during a particular time as seen from a particular place. Fra Mauro's mappamundi is a storytelling device in words and pictures, and much of that information is about comparative human behavior. The Museo Correr where the map now hangs seem to know this because they have set a bench in front of the map for those who want to sit for a while and stare, and many do. I certainly know how easy it is to sit there and become lost, or dizzy, while eyeing the inscriptions and legends even though I can't understand them. As an anthropologist I am particularly pulled to Fra Mauro's comments and judgements about human behavior across cultures, different sorts of customs and rituals and their underlying belief systems, and how his medieval Western mind took in all that "foreign" information and decided what to put on the map.

The Fra Mauro mappamundi is dotted with enumerable tiny drawings of human-made edifices such as castles, temples, shrines, walls, ships, and cathedrals, and he often discusses them, using both words and pictures to make a point. For example, eight kinds of ships sail his oceans and seas, and he includes non-Western vessels to show Westerners that other cultures use different kinds of boats than those sailing the Mediterranean and the Atlantic coasts. The drawings of Chinese junks are also explained in texts as vessels with four permanent masts as well as two additional masts that could be raised or lowered. These junks, he writes, had forty to sixty cabins for merchants, which was much larger than Western ships, and they usually had an astrologer high up on the mast who sat with an astrolabe, an ancient device that could be used for navigating by the stars, and the astrologer called out instructions from his aerie to the helmsman.[2] There is also another sleek vessel with a pointed prow and one sail coursing the Indian Ocean among the junks—this is probably meant to be an Arabic sailing vessel. Given the role of trade at the time, and his particularly Venetian perspective that trade was the most important thing in life, it's no surprise that Fra Mauro decided to showcase these boats as examples of cultural difference. These various boats are also an example of his many combinations of illustration and text to make a point, to show that the world is not just European in its habits, materials, and technological culture and that "normal" is not the same across the world.

That's why this map, with its flurry of writing and busy spread of illustrations, requires a closer look than simpler maps. It is essentially a picture book with texts of the late Middle Ages drawn from the Western perspective and it is, therefore, a window on their world.

PLACES OF INTEREST BASED ON
THE TRAVELS OF OTHERS

Fra Mauro had never visited the places he was describing. In drawing and writing about the East, he relied heavily on *Il Milione*, which was dictated by the Venetian merchant Marco Polo and written down in the late 1200s or early 1300s by Rustichello da Pisa (see also chapter 4), the travels of Venetian explorer Niccolò de' Conte, and the missionary

Odoric of Pordenone.[3] Marco Polo and Niccolò de' Conte were certainly embedded in the cultures they visited, often living there for months or years and learning the language, and participating in everyday life.[4] They were also keen observers of human looks and habits, wildlife, trade goods and where they could be found, differing languages, religious beliefs, and political actions. It's important to note that Fra Mauro is often using their exact words, or paraphrasing them, as he shortened the lengthier descriptions. Also, Fra Mauro was cherry-picking from these travelers' documents what he wanted to highlight. We will never know why he chose to textualize some parts of his map with these travelogues and why he failed to underscore others. If his lens was international trade, it would bolster his aim to visualize an international network of interconnected sea lanes, but it seems quirkier than that when comparing, for example, the rich text of *Il Milione* and the choice of specific annotated bits that he chose to put on the map.[5]

Starting with the toponym—that is, names of cities and such—some are in the language or dialect of that city or province while others are spelled phonetically. It has taken some research and work to even figure out what many of these toponyms mean and decide if they are geographically correct. To do that for the East, in particular, scholars from that area have tried to compare the place names on the map to modern names in the same area or by comparing Fra Mauro's geographic descriptions to points on a contemporary map (see chapter 8).[6] These evaluations support the fact that the travels of Marco Polo, Niccolò de' Conte, and Odoric of Pordenone were reasonable sources for Fra Mauro for Asia, Southeast Asia, and Oceana. The only question is why he didn't acknowledge those travelers at all.

In general, Fra Mauro identifies the names of what he thinks are important cities, and he often describes them. For example: "This very great city called Bisengal [probably the current city of Hampi, India] . . . has 7 rings of walls which incorporate some mountains. It has a circumference of 200 miles; and a river that flows through it divides the main part."[7] In another example, the city of Fuçui, China, is sixty miles around and has about sixty thousand bridges, which must be large and high because Fra Mauro (speaking through Marco Polo) says that galleys, large merchant ships, can pass underneath.[8] He describes the emperor of China's winter residence as twenty-four miles in circumference and having six gates.[9] Some notes on cities certainly had personal meaning. An inscription in China states, "This very noble city they call Chansay [Hangzhou] stands in a lake like Venice and has a circumference of one hundred miles, a large population, very large suburbs, and twelve main gateways. Eight miles outside these there are other cities bigger than Venice; and there are twelve thousand bridges and fourteen thousand hearths."[10]

Fra Mauro also had an appreciation of population size, which he notes frequently. Perhaps this was one way to get viewers to understand that there were many other people in the world living contentedly in different kinds of social systems. He wrote in an inscription, "And in the middle of this city [in China], there is a lake with a circumference of 30 miles, within which there are very large palaces where those that live here hold their feasts. In each house there are 12 families, which are calculated as one hearth; and these hearths total 90 *tuman* [a military term for a force of 10,000], and each *tuman* includes 10,000 hearths, which makes 900,000 hearths. And here all branches of knowledge are studied and there are magnificent things, order and abundance in all trades and crafts. . . . tuni is *tuman* which is a military term for a force of 10,000."[11] He was also enamored of emphasizing that the East was full of people although it unknown to most Europeans. He claimed that a place called Zagatai and its province in Central Asia contained 1,200 cities.[12] From these demographic statistics, people in the West were faced with the fact that, in the Middle Ages, they might not be masters of the globe.

Fra Mauro was also invested in the idea of sovereignty, again showing that places unknown to Westerners, places that heretofore might have been considered backwaters, were governmentally and politically advanced. For example, he pointed out that there were twelve cities under the rule of Nanjing, China, although it appears he got the location wrong when interpreting Marco Polo on this.[13] In his description of

Hampi, India, he wrote, "Here reigns a very powerful king, who enjoys great favour and preserves great order. To demonstrate his excellence, after once winning a great victory over his enemies and subjugating them, he offered within the temple of their city, called Turmili, four equivalents of his own weight: one in gemstones of various kinds, one in coins, one gold and one in silver."[14]

Fra Mauro was also very generous with the word *noble* and used it repeatedly for both people and cities. "For the mildness of its air, this noble island of Crete was called Macheronenson. It had noble cities and castles, and was the inventor of the oar, of bows, and of military science and of music; it also established the meter found by Pyrrhus."[15] He also thinks Sicily is very noble: "Named by the Greeks Trinachio, this very noble island of Sicily was, according to Sallus, once joined to Italy; but then the force of the sea divided them."[16] *Noble* is also the adjective for Chasay and Fuçui in China, and Sri Lanka, about which he waxes lyrical: "And this island is very fertile in all the things necessary for human life; here people live a long time thanks to the goodness of the air and the perfection of the waters."[17] Fra Mauro finds other places simply wonderful: "The island [he is referring to Scotland, which is not technically an island] is very fertile in pastures, rivers, springs and animals and all other things."[18] He pointed out "amazing" human-built features such as a bridge near Beijing that crossed the River Polisanchin: "It has three hundred arches and six thousand images of lions bearing an equal number of columns with capitals."[19] While these descriptions seem fantastic, the bridge he describes appears to be the Lu-Kou Bridge over the Yongding River, which is ten miles southwest of Beijing. Once off this bridge, he added, "There is a very pleasant road, lined for miles with gardens, palaces castles, and cities,"[20] as if he had traveled over it himself, which he hadn't. These bits of text give the map a charming air, as if Fra Mauro was a travel guide preparing his customers for a long trip to foreign ports where they would see many wonders, maybe even places that were more "noble" than their home cities.

Aside from the toponyms and the text inscriptions of cities on white Post-its on the map, many drawings of features illustrate cities, temples, tombs, and geographic features. On one spot Fra Mauro transforms a well-known mountain in China into a tower, although there is no explanation as to why he did this.[21] More often, however, he illustrated with a miniature of the place, such as a drawing of a cluster of palm trees for an oasis in Egypt.[22] In another example in Egypt, he illustrated Aswan, which was once a great city and is now best known for a huge dam on the Nile River, with a lovely miniature of the city complete with the many towers it once had.[23] Falchetta says his drawing of Hampi, India, now a World Heritage site, is even recognizable in the remains of the city we see today.[24] Fra Mauro also notes many tombs that are important in other cultures, including the tombs of emperors and their families of Cathay, which were interred on Mount Alchai,[25] another imperial tomb in China, and another in Russia,[26] and that there were many tombs on the island of Sumatra in Indonesia.[27] Some of these tombs and temples, he claimed, were also treasure chests. These sites included the aforementioned one in Cathay that held "an inestimable store of treasure,"[28] another in China where "here there is such an accumulation of treasure that one cannot even estimate it,"[29] a temple in Burma that Marco Polo previously reported as "covered in sheets of gold and silver,"[30] and a cemetery in India that supposedly had tombs full of gold and precious stones.[31] Noting these temples and tombs was, of course, also a comment about differing spiritual beliefs and funeral practices. Some might have been eager to go and find those treasures, but Fra Mauro was pointing out, using money, gold, and gems, that the world was more varied, even in death, than the Eurocentric view. As a Catholic priest, he might have also been quite interested in these different beliefs and funerary rituals in contrast to the Christian practices of buying the dead without including their wealth.

In these various notations, Fra Mauro sometimes provides local history, such as the fact that there is an ancient Roman fortress once called Babilonia (and then Masr al-Atikah) that is still inside the area of

modern-day Cairo.[32] On the map of Great Britain, he explained where
the name England supposedly came from: "Note that in ancient times
Anglia was inhabited by giants, but some Trojans who had survived
the slaughter of Troy came to this island, fought its inhabitants and
defeated them; after their prince Brutus, it was named Britannia."[33] This
note is a bit of an aside and clearly based on some previous geographer
or traveler, but it makes for an interesting tidbit for the map. Near the
Gulf of Sidra, located at the northern edge of Libya in the Mediterranean, he noted that there were "ruins of many cities, which were
clearly once very great."[34] Fra Mauro is, of course, very interested in
the Italian Peninsula (although it was not a country yet) and how
it became populated. For example, besides the story of people from
Troy, he also pointed out that the Goths from Gothia, now modern-
day Ukraine, traveled to Italy as one of the many "barbarian hoards"
that arrived on the peninsula, and that these Goths then laid siege to
Rome.[35] And he notes that the Longobards came all the way down
from Scandinavia to take over northern Italy.[36]

While describing all these wonders and all that history, Fra Mauro
also wants to warn travelers about various impediments, natural and
human-made, that might make travel difficult. Between two mountains
in Kazakhstan, for example, he noted that there was a raging, harmful
wind.[37] And on the coast of Senegal in West Africa, there was supposed
to be a famous signpost, tower, or big statue with an engraving on its left
hand warning sailors that Hercules had stopped there and, as a result,
they should go no further. But he doesn't necessarily buy this tale: "I have
often heard many say that there is a column with a hand and inscription
that informs one that one cannot go beyond this point. But here I would
like the Portuguese that sail this sea to say if what I have heard is true,
because I am not so bold as to affirm it."[38] In Russia, he claims there
were iron gates called the Caspian Gates that closed off passage through
the Caucus mountains, and historians have determined that these gates
are left over from a sixth-century defense rampart.[39] He also pointed
out other sets of iron gates in Uzbekistan, but suggested that they were

ancient fortresses that closed off a valley from trespassers.[40] There were more iron gates in Russia, but he does not explain what they guarded or enclosed.[41]

Fra Mauro also wrote a long inscription to portray two tunnels or passages used as shortcuts for Tartar caravans going to and from Cathay and tells us these tunnels are dark and one of them might have roaming lions: "It was dug entirely with stonecutter's chisel and is about twenty miles long and very dark. Thus, those inside have to shout and beat drums so that they can be heard by the other caravans coming in the opposite direction. They also do this to drive off the lions that sometimes go into the cave."[42] This detailed description comes from Marco Polo, who must have seen these tunnels, maybe even passed through them 150 years earlier. One problem with this dramatic note is that there are no lions in China or Mongolia, and there never have been. He might have been referring to the snow leopard, or this tale had been embellished with an idea of lions that someone gleaned from stories about India, where these lions are endemic.

Fra Mauro often located areas of interest with unusual animals that a visitor might see there, such as the falcons in a part of Russia, and Iran, where they make "noble works" with falcon feathers,[43] or the cranes of five different colors in China.[44] The phoenix in Saudi Arabia sounds amazing, as "the size of an eagle and has a head adorned with a crest of marvelous plumage of various colors. And around its neck, the plumage is gold coloured, whilst the wings, tail and the rest of its feathers are of purple, pink and an infinity of other colors."[45] There was also apparently a giant bird sighted by the sailors on a boat that had gone around the Cape of Diab (South Africa) and then back: "When the ship came close to shore, the sailors saw the egg of a bird that from the tip of one wing to that of the other it measured 60 feet, and it could easily lift an elephant or any other large animal, causing great harm to the inhabitants of the land; and it was very fast in flight."[46] Bird watchers today would be impressed by these specimens and those of us who know nothing about birds would also be amazed at the various colorations and adornments seen outside of Western countries.

One might expect a cartographer to include fortresses, gates, temples, and tombs on a map since they are permanent human structures. And it's normal to list populations. The wildlife is a bonus, but still, not entirely uncommon for maps of that time. What is more unexpected on this 1459 map is the sources of material portable goods that could be bartered or sold.

A WORLD OF GOODS

It's impossible to imagine any map today that might, like Fra Mauro's map, act as such an extensively combined history and geography manuscript. But that's only one intersection of knowledge on this velum. The other layer is the kinds of goods, natural and human-made, that could be found across the late medieval world. If that type of layer appeared on a map of the United States, a culture so embedded in capitalism and compelled to buy many things, we would have a map that showed where every store appears and it would also be dotted with goods from every corner of the world. Google Maps does just that when you focus in on the streets in some area of interest. Superimposed on the geography and streets, no matter the region, country, or city, there are icons and information about places to stay, where to eat, and exactly where to buy things, because this is what is currently important to our culture and our world. That modern map is an explicit advertisement, and the icons are tiny billboards that direct people where they can spend their money. Fra Mauro's map is a nascent version of the current Google shopping guide. But in his case, Mauro was also tracking exactly where to find items to bring them back to the West and sell for profit. Adding this layer would, of course, have been second nature to the Venetian monk, because buying and selling were part and parcel of his city's trading heritage. That singular economic dependency also exposed Venetian citizens to all kinds of goods from all over the world. Fra Mauro also emphasized the importance of an economy based on trade, no matter the culture or dominion, when he said of the Great Khan in an inscription, "In this port of Zaiton, the Khan keeps a large number of ships to serve his state; and it is also visited by ships from the Indias and from various areas and islands carrying different kinds of merchandise—that is, spices, gemstones and gold. For all these he collects sizable duties."[47] Fra Mauro sometimes explained the routes of trade that other countries used, such as China, where he pointed out how goods were ported river to river on their way to Cathay.[48]

Fra Mauro's map went even further than Google in supporting consumerism because it didn't just show roads that led to these sources. He also aimed to chart and connect global trade routes for bringing all that stuff back home, and that was a prescient decision. Historians have pointed out that if you step back from Fra Mauro's map and take the long view, the obvious trade routes of the late Middle Ages across water come into focus.[49] First, there were the well-known and traveled routes within the Mediterranean. One path went west, where a ship could hit any number of ports all around the shores of that sea. There was also the easterly Mediterranean route to the Levant with a spur up into the Black Sea. Although these routes are contained within a relatively small geographic space, they followed the coasts of great landmasses all around the Adriatic, the Black Sea, and the Mediterranean, which meant ships had access to any number of countries and cities. Those routes also had access to the many populated Mediterranean islands where crops were grown, goods were manufactured, and natural resources were available. Within reach of those sea routes were rivers and tributaries, even roads, available to port goods further inland. A second well-worn route headed out into the Atlantic and north to the British Isles and Scandinavia. Fra Mauro's map extended another sea route out of the Mediterranean and down the western side of Africa, where it could loop under the tip and head into the newly named Indian Ocean. From there, the Arabic and Asian trade routes were now accessible, and that meant trading directly with the East and the Spice Islands.[50] To emphasize these Eastern routes, Fra Mauro outlined on the map, in words and pictures, five established trade routes that had not been available to Europeans before, because they had no idea that Africa could be circumnavigated. He explained these new-to-Europeans routes in two inscriptions—one note near the island of Hormuz, which sits in the Sea of Hormuz between Iran and the United Arab Emirates—and the other by an island in the eastern Indian Ocean that he called Lesser Java. In these inscriptions, he described a known trade route that starts at the Chinese city Guangzhou, which is close to Taiwan, and heads to what Fra Mauro labels Chatio or Cathay, by which he meant the northern parts of China. He wrote that another such route went south from Guangzhou to the Indian coast. A third known trade route went from Hormuz, at the intersection of the Gulf of Oman and the Persian Gulf, up to Mecca on the eastern coast of the Persian Gulf, while another also started from Hormuz to the apex of the Persian Gulf and then down the Tigris and Euphrates Rivers and across land to the Black Sea, where trade was already conducted with Venetian and Genoese merchants.[51] These routes were all well-traveled by Arab, Indian, Southeast Asian, and Chinese traders. By adding a route circumnavigating Africa, the known world was now a vast network of trade routes going in all directions, and here was the new road map for Europeans. This interconnected trade route map was nothing short of an economic and world-changing miracle for the people of the West, with their unquenchable thirst for imported goods and resources and for making money.

Along with highlighting trade routes, Fra Mauro also used a remarkable number of inscriptions to name where such goods could be bought and shipped home, or what might be their sources, such as in the case of gold, silver, and gems. All this information Fra Mauro plucked from Marco Polo and Niccolò de' Conte, and although he could have ignored this level of information for his map, he emphasized it.

The most obvious trade import of interest of the late Middle Ages were spices from the East. During that time, spices were a luxury in the

West. Although some spices could be grown in the West, there wasn't right climate for planting and growing most of them. Although importing these condiments from far away was extremely expensive, Westerners were mad about small but powerful flavor additives. The import of these spices was also complicated. Venetian ships did not seek them out nor pick them up with Venetian ships—how could they when there was no water route to the Indian Ocean? Instead, Venetian traders loaded their ships up at various trading ports in the Levant and the Black Sea, where these goods had been delivered by others. Nonetheless, Venice held the reputation of spice trading for Europe because they were middlemen, shipping these precious goods from a trade outpost in the Near East to Venice and then up various rivers and roads to the rest of Europe. That meant Venetian ships full of spices routinely traveled the Mediterranean and up the Adriatic to home. The Venetian writer Andrea di Robilant wrote in his book about Venetian exploration: "It was said of Venetian ships that they left such a profusion of pungent aromas in their wake at sea that they could be detected miles away given the right wind."[52] Today we think nothing of the spice rack in our kitchen and it's no big deal when we need a teaspoon of turmeric or cardamom, which originated in the Far East or Southeast Asia, but in the Middle Ages and onward, those kinds of additives were like caviar, desired and enjoyed. From a merchant's point of view, they were better than gold because these ingredients were addictive and the consumer always wanted more.[53] All over Europe spices from the East were used for cooking, medicine, and beauty products, and they were in high demand. For Venice, spices were also an efficient moneymaking good with the perfect price point. The leaves, powders, seeds, and seed pods were small and relatively lightweight for shipping. That lowered the cost of the trade. Spices were also easy to obtain once there were established connections with traders sailing around the Indian Ocean. Quoting Marco Polo, Fra Mauro points out that people in the East had been doing this for quite some time, and they had the right boats to accomplish this kind of market. Most important, the spice trade was profitable for everyone, from suppliers to sellers.

Fra Mauro explicitly names the sources for those spices. For example, Bandan in the Indian Ocean is "a small island in the shadows on which grow a lot of cloves."[54] Today the island nation of Indonesia includes over eighteen thousand islands and incorporates large ones such as Java, Sumatra, Bali, and Kalimantan (Borneo), although it's not clear what he meant by "Java Minor." In any case, this area was the epicenter for spices, although many spices also grew on the Asian continent nearby. Fra Mauro wrote about the spice trade on the Indian Ocean part of his map: "A most fertile island, Java Minor has eight kingdoms and is surrounded by eight islands, in which grow fine spices. And on this said Java grow ginger and other noble spices in great quantity, and at the time of harvest all that grown on this and the other islands is taken to Java Major and there is divided into three parts; one part to Caiton and Cathay, another to Hormuz, Cide, and Mecca, by the Sea of India, and the third is sent northwards across the Sea of Cathay. And according to the testimony of those who sail this sea, from this island one sees the Southern Cross a yard above the horizon."[55] He also noted that "Taprobana," or Sumatra, had four kingdoms where one could find "a lot" of "pepper, cloves, aloeswood and a wood called *galambach*, which has the finest scent in the world and is sold for its own weight in gold," a comparison that underscores the value of so many of these spices.[56] Mauro also claimed that Saudi Arabia was home to other spices, including myrrh, cinnamon, and incense.[57] And there was pepper in Sri Lanka.[58] He also knew all about theriac, which was a staple in the Middle Ages that guarded against sicknesses. Theriacal recipes required several spices, meaning spices were used routinely for medicinal purposes in the West.[59] Venice was famous for its theriac and Fra Mauro suggested there was some leaf on an island in the Indian Ocean that was more "precious," possibly meaning more effective, than anything grown on continental land, but he gave no hint as to what that leaf might be.[60]

There are also several references to salt on the map. Although not what we usually think of as a spice per se, salt is a mineral additive that has always been essential for human health and it's also welcome accent in food. Also, salt was used as a food preservative long before there was refrigeration. The central position of salt in human history is why people across the globe have always been attentive to where there might be salt mountains and salt deposits, and why people all over have devised various ways to extract salt from saltwater.[61] The trade economy of Venice was initially built on the salt trade even more so than on spices. Venetians had their own salt pans around the lagoon but were smart enough to realize they could capitalize on imported and exported salt and so they demanded that incoming ships carry salt, which they taxed, bought, and then stored in huge warehouses until they shipped it out to other countries, making money at every step.[62] Fra Mauro quoted Polo when he noted salt mountains that were "the best in the world" in Afghanistan, that there was a "lot of salt collected" at a place in Russia, and that there were mountains of salt in Russia and China near Zanadu.[63]

Like a treasure map, Fra Mauro's world map also tells where a fortune hunter could find gems, valuable stones, gold, silver, and other precious metals. According to his inscriptions, there were "precious gemstones" in Central Asia, gemstones (and sugar there too) in a place in China, and rubies in Burma and Sri Lanka.[64] He also wrote that many islands in the Indian Ocean were "very rich in gold, silver and different kinds of gemstones."[65] Of Sri Lanka in particular he claimed, "The king of this island is said to have a ruby that is the most beautiful that exists in the world; a span long and as thick as an arm, it is most brilliant and rubicund and without any blemishes."[66] He also specifically mentions balas rubies from Afghanistan, information he took from both Polo and al-Bīrūnī. These large red rubies are the ones used in royal crowns and held in imperial coffers.[67] In addition, Marco Polo described the specific caves where these rubies could be dug up. India was also, and still is, another source of precious gems such as diamonds. "In this lake [In India] there is a mountain in which diamonds are to be found," Fra Mauro wrote, quoting de Conte's explorations of the East.[68] India in general, Mauro explained, was full of many kinds of gems besides diamonds—there were beryls, which include blue aquamarines, pink and orange morganites, various shades of green emeralds, green chrysoprases, and red-brown zircons called jacinths. Unspecified islands in the Indian Ocean, especially Java, also had all kinds of gemstones along with their coveted spices.[69]

Fra Mauro also pointed to some less valuable but coveted semiprecious stones such as agate: "And here [Sicily] it is said that comedy was invented and the stone called agate was first found"[70] Sadly, we can't prove that comedy was invented in Sicily because no one else has said that, but it was presumably started by Greek playwrights and scholars who lived there long ago. Given the use of lapis lazuli, with its enticing deep blue color that was set into jewelry and ground up into powder for pigments in Venice, it's no surprise that Fra Mauro remarked that it could be found in Tajikistan.[71] He also noted that turquoise could be mined Iran.[72] Fra Mauro was seemingly interested in pearls and where they were found because of place of pearls in the life of aristocratic and wealthy medieval women. Pearls were quite the fashion for wealthy Venetian women, for example, they showed up in long strands around the neck, twisted through extravagant hairstyles, and sewn into clothing as seen in Venetian portraits from the Middle Ages and the Renaissance. We know that those pearls were imported because although Venice was surrounded by water, there were no pearls in the lagoon. Fra Mauro observed, "Abapaten [in the Indian Ocean] is a small island where a great number of pearls are fished. These eastern pearls are more noble and beautiful than one finds in any other place."[73] But hunting for pearls can be tricky, he explained: "Note that in many places in these gulfs [in the Indian Ocean] at certain times one can fish pearls in great quantities and at other times none are to be found; some say this is because they pass through here."[74] And he claimed there were pearls called "Chatif pearls" to be found off the coast of Saudi Arabia.[75] This statement might refer to the variety of gulf pearls that come out of the Persian Gulf even today.

Mauro also noted a source of amber, which is neither a stone nor a gem but fossilized tree resin that ranges in color from bright yellow to golden brown; sometimes insects or bits of plants are caught in the fossilization process, and they become a decorative feature within a blob of amber. Today, amber, like all the aforementioned gems, continues to be a popular material for jewelry and larger objects such as pipes, book-ends, and lamps. Amber is found all over, but Fra Mauro pointed to one source in the Indian Ocean: "Mahal, an island inhabited by Christians, where there is an archbishopric. On said island there is a lot of amber to be found, and the pirates of these seas have their stronghold here, and here they sell and store their loot."[76] Historian Piero Falchetta says that Fra Mauro is referring to the Maldives Islands, which lie southwest of Sri Lanka.

Fra Mauro's notes on where to find gold are extensive and geographi-cally wide-ranging. According to him, gold could be found in Africa, China, South Africa, Central Africa, Senegal in West Africa, Djibouti in the Horn of Africa, as well as on the islands of Sri Lanka and Java.[77] Most of these sources have been confirmed, but these many citations also point to a gold fever that had infused international trade, the same fever that sent ships down the west coast of Africa and later across the Atlantic in search of this precious metal. It's no wonder he even claimed, "Here there is gold dust" in one spot in China, although it's hard to know what he meant by gold "dust."[78] Fra Mauro also remarked that there might be silver deposits near a pass between Kazakhstan and China and that many islands in Oceania also have gold and silver.[79] Less valuable metals such as iron and copper, among others not specified, could be found on islands in the East China Sea, in Saudi Arabia, Iran, and Burma, where "a city had walls of copper that are a yard thick."[80]

Surprisingly, Fra Mauro said little about the types of cloth that could be found in foreign places. This is odd because fine cloth, especially silk, was a major trade item for Venice. But he does say that in Azerbaijan three types of silk were produced and that a "great quantity" of gold cloth was made in the now lost or renamed city of Paugin, China.[81] Addressing another luxury item, Mauro reported that a specific city in China made porcelain. Those Chinese porcelains were fine works of art and they were known and coveted by Europeans.[82] Interestingly, he noted shipments of leather, which is, of course, cured animal hide. Animal skins were mentioned in Russia, where one could have traded for pelts of ermine, sable, and other exotic animals from the northern climes, and this kind of trade was well-known in Fra Mauro's time.[83] He also noted there were leather hides on some Catalan (Spanish) ships traveling through the Barents Sea, off the north coast of Russia.[84] Again, leather clothing and ermine and sable pelts can be seen in portraits of the European aristocracy from the Middle Ages and the Renaissance. And hides, such as beaver, became desired trade goods from North America after it was "discovered." Europe also had a thriving market in leather made from endemic animals such as sheep, goats, and calves, but exotic ones like ermine had to be imported.

Although we don't think of crops as trade goods because of possible spoilage, especially on long sea voyages, they were times when those crops might be valuable or needed across continents. Today, grains—including wheat, sorghum, and teff, among many others—as well as beans and soybeans, and of course mountains of refined sugar or sugarcane cross continents.[85] The same kind of bulk trade was common in the Middle Ages. For example, Fra Mauro pointed out that there was sugar in China, which was probably sugarcane, peppers in India that could travel dried, and cotton and corn growing in Iran.[86] Not all large-scale exportable crops were foodstuffs. For example, he mentioned the scrub indigo found in Sri Lanka and Iran, which would have been an important import for Europe because it was used for dying cloth and making ink and oil paints blue, a prized color for clothing and the arts.[87] Quaintly, he made a major mistake about the use of Egyptian pyramids and does not know they are tombs but instead calls them "granaries of the pharaohs." Although amusing now, before the pyramids had been opened it was a logical guess.[88]

More fun are Fra Mauro's texts about particular items. Read his description of the durian fruit of Sri Lanka, a fruit that is either loved for its taste or hated for its smell throughout Southeast Asia: "Here trees bear a fruit called the *durian*, which is the size of a reasonably big watermelon and has a green, knobbly skin, rather like a pine cone. The five fruits inside is each one the size of a reasonably big pine cone and each one of them has its own pleasant taste; inside they are reddish-purple in color and they are very warm."[89] And he certainly puts a finger on one of the best fruit resources of Morocco: "In this desert there are date trees—and beyond them no more are to be found."[90]

Fra Mauro also often pointed out various natural resources that might be valuable, or just interesting.[91] He mentions ambergris a few times, for example. Ambergris is a material produced in the intestines of ocean-going whales and it might be an important lubricant in the workings of a whale's digestive system. Ambergris is passed out of the whale's body along with digested food and it ends up floating on the water or washed up on beaches in gray lumps. Back in the Middle Ages, it was used as an ingredient in perfumes and perhaps incense because it has a musky odor. Mauro also commented about some great honey from Africa and passed along a story that might or might not be true: "In this wood of Abassia [Africa south of Egypt] there is such a great quantity of honey that they do not bother to collect it. When in the winter the great rains wash these trees, that honey flows into some nearby lakes and, thanks to the action of the sun, that water becomes like a wine, and the people of the place drink it in place of wine."[92] The same sort of language is used to write about some sort of liquid from Azerbaijan that supposedly had medicinal properties: "Toward the coast in this province of Siroan and Siamachi there are two 'liquor' springs. From one, the larger flows a green liquid called *nephto* [naphtha]; this is good to burn and is very common in Syria and Asia Minor. The other liquid is white and is medicinal; it is good for various things."[93] He also described manna, that mysterious substance that was thought a life-sustaining gift from God, which he claimed formed nicely on leaves in Saudi Arabia. Other manna that formed on rock was, according to his sources, not as good.[94] Given that many different foodstuffs have been called manna over history, who knows what he was alluding to here. In the same vague way, Fra Mauro wrote of fruits, timber, herbs, and roots "with virtuous properties" in India.[95]

While not described as trade goods for the West per se, Fra Mauro often waxed lyrical about various animals in other countries. There were

"noble horses" and "highly prized mules" in Iran that are sold at markets in India.[96] In Sri Lanka, he claimed, there were "completely black
lions and white parrots with red beaks and claws" and "large numbers
of elephants in this place."[97] There are only leopards and some small
wild cats in Sri Lanka today, so it's hard to know what the black lions
might have been. He noted that on the island of Bandan in the Indian
Ocean, the parrots were completely red, except they had yellow beaks
and claws.[98] In a sort of natural history lesson, he repeated from some
source that in the Sea of India, "there are some fish which, if pursued by
other larger fish, enter into the body of their mother; once the danger is
over, she opens her mouth and they come out."[99] In Norway there were
many animals unfamiliar to most Westerns in the Middle Ages; he tells
of especially "huge white bears and other savage animals."[100] Perhaps
to Fra Mauro, these were wonders and so they deserved a place on the
map. Today we just think of them as part of biodiversity, and we expect
to hear constantly about heretofore unknown species as humanity tries to
record and save the rich biome of life on Earth while it is disappearing.

The extensive list of items on Fra Mauro's map is surprising and reve-
latory because no other map had ever attempted to plot material goods,
let alone animals and crops, geographically. Of course, it made sense
given that Fra Mauro was Venetian, but still, the decision to accord these
goods a definitive place in the world was telling. This period in human
history was the beginning of the Age of Consumerism and that economic
strategy began through global trade. In that sense, Fra Mauro's map was a
portent of the future. Today we track closely the price of crude oil, which
comes from various continents and countries, and when and why that
supply is in danger, as it was so graphically during the COVID-19 pandemic when countries shut down and global supply chains were broken.
And with the invasion of Ukraine by Russia in the February of 2022,
everyone halfway across the world held their collective breath watching
the rise of the price of crude oil that is now sent around the world to
supply energy for vehicles, factories, and home heat. Today, highlighted
by the war in Ukraine, oil companies are involved in territorial disputes

and hold the fate of nations in their supply chains as they become political actors in a war on another side of the world. This situation is the very definition of international trade and how it operates, and how it can be broken. Imagine if the Spice Islands had gone to war with, say, India in the Middle Ages and cut off the supply of spices to Europe as a result, or if other places cut off the supply of salt. Modern peoples are just as dependent on or addicted to trade goods as were people in the West in the late Middle Ages. All those boats bobbing around on Fra Mauro's map are representations of the many cargo ships and planes that crisscross the world today transporting goods here and there for profit. At this point in human history, we are not only joined together as one species, but we are also joined together as one customer, dependent on the goods of others to have the lives we are now used to. Oil today is the spices and salt of yesteryear, and that dependence has turned us into a species that relies on the goods found by, grown by, or manufactured by a host of unknown "others" who live so very far away.

A PLETHORA OF PEOPLE AND
THEIR CULTURAL PRACTICES

Fra Mauro was a man of Europe, but since he was living in a place that was the center of global trade, he was witness to a vast array of people from various cultures from around the world. That heritage, and the well-known works of Venetian explorers, set him up to think about, and then decide to catalog, the rainbow of peoples and cultures on his map. We don't know anything about Mauro's original notes on these cultures because those personal pages were lost long ago, but we can consider this map the first visual display of human cultures around the world from the viewpoint of the Middle Ages in the West. Of course, not every culture shows up on the map which is not a surprise; even today we have no idea how many cultures are still hidden from Western view. But today, the knowledge we do have about the variety of human cultures also demonstrates that that not every place on Earth has been Westernized into one giant group with the same practices and thoughts.[101] One of the interesting features of the human species is that we group together and share beliefs and play them out, and we can stubbornly hold onto that lifestyle forever, or we can move to another place and shed that suite of beliefs and behaviors and take on the lifeways of another culture. So, we might be varied in our practices and beliefs, but we are also a very flexible species in terms of culture.

In some ways, Fra Mauro was an anthropologist.[102] For example, he named six tribes or cultures in India and said, "Those who want information of their very different customs and practices should read Arrianus ad Strabo, who write extensively about this India."[103] He also went to great lengths to name specific peoples such as the Muritaninas, Xavi, Permians, Scots, HmrHmr (Berbers), Benichileb, and Aragafi, among many others. And he held some of these cultures, or their rulers, up for admiration. He also loved and did not question stories about exaggerated pomp and circumstance: "This most excellent and mighty emperor [the great khan living near Beijing, China] has sixty crowned kings under this dominion. When he travels, he sits in a carriage of gold and ivory decorated with gemstones of inestimable price. And this carriage is drawn by a white elephant. The four most noble kings of his dominions stand one at each corner of this carriage to escort it; and all the others walk ahead, with a large number of armed men both before and behind. And here are all the genteel pleasures and customs of the world."[104] Of someplace in India he noted, "As well, there are cities, castles and innumerable people of different varieties, standing, and customs. There are powerful lords, great number of elephants and a diversity of almost incredible monsters, both human and animals. For example, there are serpents and other horrible beasts—especially the *euchrota*, the fastest of all animals."[105]

Fra Mauro also came close to being an anthropological linguist when he recognized that the name of a river in Asia changed as it passed through territory after territory and the native languages changed accordingly.[106] Across the Caucasus Mountains, he noted that the names of the mountains also changed "because of the diversity of languages of the people who live up there."[107] He further commented, "In these mountains [of Georgia] there are said to be more than thirty different languages, and many religious faiths and even more different customs."[108] In making those observations about languages and customs, Fra Mauro was also recognizing the variety and dynamic fluidity of cultures, and that some could be better, or more interesting than a Westerner might expect. He wrote of a city in Iran, "Here they practice all the everyday crafts and there is study of every kind."[109] That kind of recognition and respect for other cultures was also displayed when he used his favorite word, *noble*, or highlighted their place on the map and said complimentary things. The city of Thasi in Yemen is called "a place of great state, justice and liberty, and all kinds of foreign people live here safely,"[110] and when he lauded a "great university" in Spain.

Fra Mauro even thinks some places are better to live in than Europe, and that the people are wonderful over there. In Sri Lanka, he tells us, there are lots of elephants and, "the people of this island live in better conditions than those of the nearby islands; they are well-formed, strong, polite and good astrologers. They are also taller than those who are born in India, just as their elephants are bigger than those in India, and those in India are bigger than those in the Mauritanias. This is due to its fine location and air."[111]

Fra Mauro's approach to cross-cultural religion and spiritual beliefs was obviously skewed by his devotion to the Catholic faith, but there is less fear of other religions on this map than on other medieval maps. Nonetheless, he paid heed to traditional Christian highlights. For example, he puts an inscription and illustration of Noah's ark next to Mount Ararat in Turkey.[112] But his forte was commenting about religions in other places and his many notions about idolaters and pagans. "Similarly, in this Persia [Iran] was the magical art discovered; and here, after the confusion of languages of mankind, came Nembrot the Giant, who taught the Persians to adore the sun and Iran fire. They call the sun *Hel*, but now they are largely Muslims. Some of them adore idols of different ways and with a great variety of faiths."[113] Fra Mauro also made a point of identifying places where certain kinds of religious or spiritual practices were different or similar to Christianity, such as "in this province [in Myanmar] there are large numbers of pagan hermits,"[114] or "on this mount [in Ethiopia] is a lake and abbeys of holy monks."[115] His notations of various funeral practices also underscore his interest in varied spiritual beliefs. People on Java burned dead bodies to ashes,[116] in India Brahmins put the ashes of the dead into golden jars and tossed them into lakes so deep that they would never be seen again,[117] and at one place in Russia, only eighteen chiefs of the Asian conqueror Timur were allowed to be buried.[118]

Fra Mauro's perspective as a religious person, Christian, and a Roman Catholic is surely behind his many attempts to explain where all the Christians were, but he did not make many judgments about this information. Perhaps his marking the Christians across his map was meant to reassure viewers that there were Christians all over the world. One inscription reads, "Note that there are two Sayts [in Egypt]: in the Upper one there are black Christians, and in the Lower white,"[119] and that there were Christians in Algeria.[120] In other places, Fra Mauro used words that were symbolic of Christian territories that had been surrounded by pagan or Islamic populations.[121] According to him, there were two islands in the Indian Ocean, Nebila and Mangla, which were occupied by Christians, and so Christian that the women stayed on one island and men on the other; mixing was only allowed for three months of the year.[122] Outside the Christian belief system, he identified other religions across the map, and disparaged many of them. There were some kind of "idolaters" on Sumatra,[123] "savage, untamed idolaters" on three other islands in the Indian Ocean,[124] idolaters on the Andaman Islands in the Indian Ocean,[125] and "evil men who were not Christian" in northern Iceland.[126] Sri Lanka seems to have people of many religions, including Jews, Muslims, and "idolaters."[127]

Fra Mauro also had a postmedieval curiosity about how people made a living. Of some groups in Central Asia near the Caspian Sea, he said,

"They live up in the mountains, where the inhabitants—or, at least, most of them—work iron and make weapons and all that is necessary for the military art."[128] He also had an interest in how businesses operated in other places: "Their custom is to come at a specific time to the place set aside for this trade. Here, they put alongside the salt the amount of gold they think fit, then they leave. A day later they come back, and if they do not find the gold then the salt is theirs. If the gold is still there, then they add what they think fit. And they go on in this way until a bargain is struck, and the parties to the trade neither see nor speak to each other," which is a description of barter by a group in West Africa.[129] In contrast, the people of the Maldives had it pretty easy: "The people on one of the islands of the Maldives in the Indian Ocean had an 'abundance' of Venus shells picked up on their beaches which were used for money."[130] He was also interested in what people lived in: "These Permians [in Russia] lived further to the north than any other people; they make their houses underground because of the great cold there is in winter."[131]

Fra Mauro often commented on the personalities or attitudes of various peoples, sometimes judging their behavior in a bad or good light, which of course does not follow the tenants of cultural relativity, where such moral judgments are avoided. But scary negative comments about different cultures are rampant in Marco Polo's book and he is Fra Mauro's biggest source. For example, Polo explains, "The men wear earrings and brooches of gold and silver set with stones and pearls. They are a pestilent people and crafty; and they live upon flesh and rice," which is an interesting observation followed by a negative opinion. And so it's no surprise that Fra Mauro, who plagiarized extensively from Polo and de' Conte without acknowledging them, is often repeating some of that.[132] For further examples, Fra Mauro claimed that "the people [of Scotland] are of easy morals and are fierce and cruel against their enemies; they prefer death to servitude."[133] Norwegians were "strong, robust and of great stature," and the men also very fierce, so fierce that "according to some Julius Caesar was not eager to face them in battle," and these people were "a great affliction to Europe," and even the Greeks could not dominate them. He also added, "But now they are much diminished and do not have the reputation they formerly had."[134] Other cultures, he felt, had been corrupted over time. "In ancient days here [in Afghanistan] ruled a lord known as The Old Man of the Mountains, who through his cunning had created a place full of every delight and pleasure. Here he brought men and fooled them into thinking it was paradise. Out of devotion to this lord, these men then committed great robbery and murder, which was the reason why a Tartar seized control of this place."[135] The people of Lithuania, the Sami, "are men of poor condition and standing," so he says.[136] And oh those Permians in northernmost Russia: "These Permians are the last people to the north of the inhabited world. They are tall, fair-skinned, strong and brave; but not industrious. They live on wild game and wear animal hides; they are men of bestial habits, and to the very far north they live in caves and underground because of the cold."[137] According to Fra Mauro, Permians were also "near savages" because they ate sable and ermine and then wore their pelts, which is pretty standard behavior for Arctic hunting cultures. Think Inuit and seals.[138] In West Africa he noted that the Benichile area was inhabited by "very strong and great people who live in great fortresses on the massive waters of rivers and on mountains. These people have dog-like faces and could not be subjugated by the Romans."[139] He reported information from Niccolò de' Conte that some tribes of the Andaman Islands off the coast of India were cannibals, which has neither been confirmed nor denied in modern times.[140] He also told of the "Giantrophagi" in Western Africa who supposedly ate human flesh.[141] Other groups were scary but contained and so they were considered better fellows. In North Africa "there is a very savage and idolatrous people who are separated from Abbassia [Africa below Ethiopia] by a river and by mountains, at the passes of which the kings of Abbassia [on of which he considered to be the mythical Prester John] have built a great fortress so that these peoples cannot pass and do no harm to their country."[142]

In total, the layer of this map that focuses on people and what they do takes it beyond all previous maps and elevates Fra Mauro's version into the realm of an encyclopedia. No matter that he never visited these places himself, no matter that his information was, by and large, provided by three travelers and possibly some sailors from other continents and cultures who passed through Venice. The point is that Fra Mauro thought this information was important enough that it should be a central feature of a world map. All these inscriptions about other ways and other people were deeply intentional. Fra Mauro was his own editor, and no one was telling him what or what not to put on this map. It was made in private at a monastery island near a small city in Italy, probably with little supervision from his religious superiors. In that sense, these inscriptions elevate this map to a new level, making it a knowledge base for the ages, a document of human behavior during one period of history, and therefore a window to the past.

MONSTERS, MIRACLES, MYTHS, AND MARVELS

One of the signature features of most medieval maps is the appearance of mythical, often dangerous, creatures. Fra Mauro has some of these, too, but not to the extent of other maps of the age, which are crowded with sea monsters and angels. Those illustrations are, of course, significant. Some are there as warnings, others to point the way to heaven; most often they act as propaganda to reinforce a religion or the politics of a kingdom.[143]

Fra Mauro also had something to say about many of these creatures. For example, according to him, there were dragons in the mountains of Afghanistan that had special stones set into their foreheads that "cures many infirmities," but it was possible to kill the dragon and pluck out the stone. "When those of the place want to kill dragons, they start great fires in the woods that are around the mountains, so that the smoke is so thick it suffocates them. When they are dead, they break open the forehead and find the above-said stone; and with that flesh—mixed with other medicines—they make a theriac [the all-purpose medicine of Europe] that is very good for many infirmities.[144] At the same time, Fra Mauro didn't necessarily believe all this hokum about dragons and beasts, which put his map on the road to scientifically based cartography. At one spot he wrote: "Here [in Russia] there are said to be a lot of monsters, which I do not give because they are almost incredible."[145] He held the same thought about curious stories from Africa: "Because there are many cosmographers and most learned men who write that in this Africa—and above all in the Mauritanias—there are human and animal monsters, I think it necessary to give my opinion. Not because I want to contradict the authority of these men but because of the care I have taken in all these years in studying all possible information concerning Africa," after which he lists the various stories and says he could never find anyone to confirm them. "Thus, not knowing anything, I cannot bear witness to anything; and I leave research in this matter to those who are curious about such things."[146] Fra Mauro took the same tact with monsters reportedly on the islands of the Indian Ocean: "Some write that in these Indies there are many types of human and animal monsters, but because few people believe these things, here I have made no note of them, except for certain animals, such as the serpents which are said to have seven heads. Again, here there are ants so very large that—something that I will not dare to say—they seem to be dogs. These could be a species of animals that are similar to ants."[147] The last part of the sentence shows Fra Mauro's scientist mind, thinking that maybe these creatures could look like something but on closer examination might be normal creatures, turning a fantasy into reality. That skepticism about scary monsters allowed him to repeat only one sea legend at the far north of the map. "These two gulfs [probably in the Barents Sea] are very dangerous for sailors because of certain fish, which puncture the ships with a spike they have on their backs. There is also another sort, like eels, which have a beak that is as hard as iron and can pass through any wood. Thus those who sail here keep close to the coast to avoid danger. And I have this from men worthy of credence."[148] Presumably hearing this tale from sailors made it more real for him and worthy of note on his map. Historian of monsters on medieval maps Chet van Duzer says that during the medieval period, Fra Mauro was the only cartographer who was skeptical about the existence of such monsters. [149]

Alchemists of the day were devoted to a scheme that claimed to turn inert items, such as iron or wood, into gold after dipping in some sacred water source. Medieval writing and maps repeated these tales and explained where this might happen. Fra Mauro has a few of these as well, but his texts usually added a soupcon of skepticism: "In the island of Hibernia [Ireland], which is most extraordinarily fertile, it is said that there is a water in which, if you immerse wood, after a while that part of the wood which is in the earth becomes iron, whilst that in the water becomes stone, and that above the water remains wood. And if one believes this thing, one can also believe in the lake of Andaman." Here he is referring to the Andaman Islands in the Indian Ocean, where legend also claimed water could turn wood into gold. [150] He then recommends a reading list of classical writers for support and ends with, "Similarly they can read Aristotle's Meteorology and Pliny on the wonders of the world, and they will see thousands of things of which I have not mentioned one." [151]

Fra Mauro was telling the viewer that there are many folktales like this floating around, he has investigated some of them, and it's just not worth the effort to repeat them or draw them on the map. And yet, a few fanciful tales slipped through: "The inhabitants [of the island of Socotra in the Indian Ocean] used to be generally Christians, and they are necromancers; through their art they deny or sell sailors favorable winds." [152] He heard from historians of a place in West Africa that is "so hot at night that anyone putting their hand in the water would be scalded; whereas during the day, the water is so cold one cannot stand it." There were "bestial customs there and monstrous animals—such as serpents, dragons and basilisks—and give other information I cannot mention here" in West Africa as well, so he's heard. [153] Fra Mauro is so hesitant with these tales that it's more likely that thinks these accounts were untrue, and he certainly did not trust those so-called historians.

And yet, Fra Mauro also imparted some long stories on the map that seem to have been handed down over the centuries, and this first one, about the island of Sri Lanka near India, is connected to Adam, of Adam and Eve. Perhaps his connection to Christianity made the story more real to the monk, and therefore more worthy of his map. "It is said that in this island there is a mountain named after Adam, the summit of which is so high that it never rains there, nor does one feel the wind. To shorten the route to that summit, you ascend by the six iron chains fixed end-to-end into the mountain by order of Alexander the Great. And on this summit it is said that the stone bears the print of Adam's right foot, in which appears many rubies; the inhabitants say that Adam did pass by this mountain."[154] This note suggests that after Adam was expelled from the Garden of Eden, he ended up in Sri Lanka, a story that seems unreal to the modern mind but perhaps made sense in medieval times.

Western medieval maps always had some depiction of people, or tribes referred to as Gog and Magog.[155] They also show up in the Koran and the Hebrew Bible. They are part of Christian doctrine, where they are known as two groups of barbarian people, or sometimes described as one savage people called Gog living in a land named Magog.[156] In all cases, these words meant danger, a darkness, because these savages, singly and as hoards, were intent on ravaging the land and killing, and then eating, everyone in their path. Satan was involved here, as well as Jesus who was supposed to save Christians from the annihilation brought on by these beasts. The Romans added to the fable the idea that everyone had been rescued from these savage hoards because Alexander the Great had closed them off from the rest of humanity and installed iron gates to keep them in, if only those gates held fast.

Over time, stories of Gog and Magog were ubiquitous, pinning them all over the place. These beastly barbarians were hiding or constrained by walls and gates in what were then known as the Caspian Mountains, the very far north of Russia, near the Great Wall of China, or the northeastern region of Asia. Or they were in the Middle East, Central Asia, Eastern Europe, Mongolia, or wandering about Hungary. They have also been called the tribes of Israel. In other words, the legend of Gog and Magog has been a catch-all phrase for scary people who might swoop down and kill everyone and take over. This legend lives on into today; it's an oft-repeated conspiracy theory morphed and repeated on the threads of the internet, even though we can now see everywhere on Earth with Google satellite maps and photographs and video from outer space. No matter. Some people right now are stuck in medieval thinking, and they are still afraid of the unknown, worried about what goes on in unfamiliar (to them) places. Perhaps it's just human nature to be afraid of the unknown, and sometimes people make up a place or people to hold those fears, such as Gog and Magog, as a paranoid catch-all.

Given that Fra Mauro was a man intellectually bending toward science, it's somewhat surprising that he, too, had Gog and Magog on his map, but those identifications were, at the same time, discounted. For example, one inscription in China reads: "Here it is said that these people were enclosed by Alexander in these lands of Hung and Mongul and that they derive their name from those two lands, which amongst us are called Gog and Magog. But I do not believe this opinion."[157] He has a much longer discussion about Gog and Magog in a text placed near Mount Caspian: "Some write that on the slopes of Mount Caspian, or not far from there, live those peoples who, as one reads, were shut in by Alexander the Macedonian [the Great]. But this opinion is certainly and clearly mistaken and cannot be upheld in any way because the diversity of the people who live around that mountain would certainly have been noticed; it is not possible that such a large number of people should have remained unknown given that these regions are fairly well-known to us."[158] Good point. If a bunch of savage hoards had been enclosed with walls and a giant gate, surely everyone in the area would know about it. He also pointed out that in that very busy part of the Caucasus Mountains, which was an established Asian trade route, there were Georgians, Mingrelians, Armenians, Circassian, and Tartars, among many others, traveling that route. Then he relocated Gog and Magog elsewhere. "Very far from Mount Caspian and [they] are, as I said, at the extreme limit of the world, between north-east and the north [meaning northeast Asia], and they are enclosed by craggy mountains and ocean on three sides." Fra Mauro then renames them Ung and Mongul. Even more interesting, he goes on to discount the common Christian prophecy that the Gog and Magog hoards will be unleashed "at the time of the Antichrist."

All this moving about of Gog and Magog might just be a result of expanding cartography. Historian of the map Piero Falchetta points out that the area of Mount Caspian used to be considered the easternmost part of the world on maps. In other words, for a very long time it was the edge of the known world and so a good place to locate unknown, threatening, apocalyptic hoards. But with the expansion of world maps to include the complete Asian landmass, they could be pushed even farther out. In the end, Fra Mauro agrees with St. Augustine, following the teachings of St. John, who had said that bad people are not one group off in some remote corner of the world but spread all over, which is a pretty good explanation of how there are bad people everywhere. In the notes on the map, Gog and Magog are transformed from some awful club-wielding tribe waiting to wreak havoc across the earth to what we all know is true—there are bad people all around us.[159]

The other long-held Christian myth that appeared on most medieval maps was the legend of the Prester John, who was held up as a leader, a patriarch, of the Eastern (Nestorian) Christian religion.[160] He was also sometimes considered a descendant of the three magi who visited baby Jesus in the manger, which gave him significant credence as a holy man. The title Prester means "a priest of high status, one who is revered." As the story goes, there was a man in Mesopotamia or thereabouts who was a priest and he wanted to convert pagans and Islamic peoples to the eastern version of Christianity. The priest was so successful at this task that he gathered many converts around him and became a leader, a king of sorts, with an army of followers and a cache of gold and jewels. This legend appeared in the twelfth century and was incorporated into Christian thinking for the next three hundred years, right up to the time of Fra Mauro's map. In the legend, as it was handed down, Prester John and his followers were initially located in India, sometimes in Central Asia, and then by the mid-1400s he had somehow landed in Ethiopia to defend Christians there against Islam and convert the Muslims to Christianity. The kingdom of Ethiopia had been Christian since the fourth century and Prester John, as protector of African Christians, was called the "King of Africa" by Westerners, and he was an established part of African Christian mythology.

Fra Mauro believed in the tales of Prester John, although he did not locate the so-called kingdom of the priest-king with a special icon or illustration. He mentioned John as the ruler of "Abbassia," the area of Africa that included everything below Egypt, but Westerners at that time had no idea that Africa was gigantic below the Gulf of Guinea. Fra Mauro wrote in an inscription: "Above the kingdom of Abbassia there is a very savage and idolatrous people who are separated from Abbassia by a river and by mountains, at the passes of which the kings of Abbassia [Prester John was apparently only one] have built great fortresses so that these peoples cannot pass and do harm to their country. These men are very strong and of great stature and they pay tribute to Prester John, King of Abbassia, and two thousand of these men serve him to his needs."[161] Fra Mauro also said in an inscription placed on present-day Ethiopia, "Here Prester John has his main residence."[162] In another inscription nearby he gives a longer description of Prester John's supposed kingdom: "It is said that Prester John has more than 120 kingdoms under his dominion in which there are more than 60 different languages. And all of this number—that is, the 120—it is said that 72 are powerful seignories [governments], and the others are not of much account."[163] Furthermore, "The king of Abbassia, called Prester John, has many kingdoms under his dominion; and his enormous power is held in esteem because of the numbers of his people, who are almost infinite. And when this lord travels with his armies, he has with him one million men, who go naked into battle, except that many of them wear crocodile skin in place of armour."[164] There is, in fact, no documentation of a real Prester John, nor any evidence of his so-called kingdom. And yet, the Ethiopian Coptic Christian delegation of priests that had traveled to Florence in 1439 to attend the meetings about combining Roman Catholic and Eastern Orthodox faiths, as well as the Coptic priests who may have visited Fra Mauro in Venice, were also part of this history, passing it along as they traveled. The tale of Prester John was not one about barbarian hoards taking over, like the story of Gog and Magog, but a fantasy about religion and the need to defend the faith and covert others before being taken over by infidels.

All this business about Prester John was probably a challenge for Fra Mauro. Historian Angelo Cattaneo says that he was up against stories and texts that were considered true by traditional authorities.[165] But he had no first-person sightings, no real evidence. And there he was, on the waning cusp of the Middle Ages, not yet in the enlightened Renaissance where stories of Prester John might have been taken more with a grain of salt or checked out across several sources and various platforms. He was also lacking the maps that came after his, those that would be drawn with much great information and accuracy during the Age of Exploration. An extra century would have also made the African continent more cartographically accurate and African kingdoms, tribes, and religions more realistic. With all that information, Fra Mauro might have been able to ferret out the truth of Prester John. On the other hand, as we know too well today, myths can be debunked, but that does not mean they go away. The wacky conspiracy theories that inhabit our politics today are all based on previous tales and fables like Prester John that, in the end, have no basis in reality. All these stories do is provide some odd sort of comfort to very scared people of certain faiths who are threatened by other religions, other people, and other ideas. Prester John was the QAnon of his times.

And so, on Fra Mauro's map there is a mixture of what might be called "science-based" information, geography, and text that could be checked out and substantiated by several people as well as nods to other bits of information that we now know is bunk. Fra Mauro was an arbiter trying to decide which among his sources were correct and which were worthy of criticism. The very fact that he was working through this process, and that we see him do it on the map, signals a shift from faith-based maps guided blindly by the hand of the church to an individual cartographer, a scientist really, making decisions about what to include, and, therefore, what he wanted to share with those who viewed and studied his map. In that sense, Fra Mauro's map led the way to all world maps that came after him.

This map, virtually abandoned for centuries on an island monastery in the Venetian Lagoon and now hanging quietly on a wall in a museum in Venice, was the cartographic and anthropological spark that eventually electrified the human vision of the world as it would be shown on all the world maps that came after.

CHAPTER 7

The Consequences of
Fra Mauro's Map

*The world map is a translation in the literal sense
of the word; it carries knowledge over, across
space and time and between cultures.*
— Marianne O'Doherty, "Fra Mauro's World Map"
(c. 1448–1459), 2011

*Magellan's voyage [1519–1522] proved, once and for all,
that the Earth was round and could be sailed around.
The truce scope of the Earth began to take shape in the
human mind, vaguely at first, then more surely, much
as the solar system is being revealed to us today.*
— John Noble Wilford, *The Mapmakers*, 1981

*Maps could easily be convicted of perjury. They do not tell
the truth, the whole truth, and nothing but the truth, for the
simple reason that the truth is not in them. Their very nature
makes them liars, and even those maps drawn by the best
cartographers from the most reliable data are falsifiers.*
— Otis P. Starkey, Professor of Geography,
University of Pennsylvania, *New York Times*, 1942

The late Middle Ages was an exciting time for Western culture. The West was just on the cusp of breaking out of its known geography and sailing to far-flung places, discovering the rest of the world, and making connections across the globe. But this Age of Discovery (or Age of Exploration) was not so much about finding new places for the fun of it. Instead, all that seemingly noble exploration and so-called innate quest for human knowledge was primarily a financial move. When Europeans started moving far beyond their geographic comfort zone, they were incited by capitalism—that is, the desire to pick up goods and resources from foreign countries and sell them back home, or elsewhere, at a profit. Fra Mauro's map was both pivotal for ushering in those capitalistic changes and reflective of the various intellectual revolutions that had started to flower. The following decades gave birth to the Renaissance, an era in Europe that rediscovered the Greek philosophers, upheld humanistic values, encouraged artistic expression, and formulated the scientific method. Like no other world map before it, this one was brimming over with information from other places and cultures, suggesting there was a wide world out there (or at least half of it) and that people in other places did not think or act just like Europeans.[1] It hinted that there was so much more to learn and understand. And that makes it a reflection of the tipping point that brought Western culture out of the Dark Ages into the light of modernity.

The role of Fra Mauro's map in encouraging European expansion and knowledge can be seen in the long series of world maps that came after. This medieval map also, in a way, instigated various attempts to take a round globe and flatten it out for easier viewing and figuring out how navigators could crisscross the world as its expansiveness became more real. It also encouraged reality-based geography and played a role in the beginnings of the science of the history of cartography. In other words, Fra Mauro's map was like a pebble thrown into a pond, creating various unpredictable but sizable waves that spread out from the initial impact; it changed world history, and how world maps have since been used for various purposes, and the discipline of cartography. That's why Fra Mauro's map is not just a map—it is a flexion point.

So far, we have looked back on this map, but now, we look forward.

THE COPIES

If this map had such a strong influence on cartography and exploration, it should have been a smash hit at the time, viewed and studied by hundreds. But that expectation is far from the truth. Of course, the late Middle Ages was not the digital age, and there was no practical way that a map hung in a monastery on a Venetian island could have had instant fame. But there was interest. We know this because of the history of various copies of the map that were made over the next many centuries, and from hints in other documents that some people used it as a reference for their own purposes. That consistent, ongoing murmur of attention that lasted for centuries demonstrates that Fra Mauro's map was known and appreciated by those for whom world cartography was an obsession, and it underscores the idea that this map was a driving force behind the progressive move into science-based mapmaking as well as the eventual establishment of the academic disciple of the history of cartography.[2]

King Afonso V of Portugal commissioned the first copy of the map in 1457, when Fra Mauro was still alive. Fra Mauro and Andrea Bianco made this copy, and it took them two years. As documents in the monastery records show, it was a paid commission and was sent to Lisbon in the spring of 1459. It makes perfect sense that a Portuguese ruler would want to have a good look at Fra Mauro's world map. Portugal was becoming a nation dedicated to exploration; they began that exploration by leaving the Mediterranean and taking a left down the west coast of Africa. Also, King Afonso V's nephew was Prince Henry the Navigator who spent his life encouraging mapping and exploration, all with an eye toward profiteering. Prince Henry died in 1460, but he might have seen the map copy after it arrived in Portugal in 1459.[3] The Portuguese were, at that time, considered master sailors and they were responsible for Fra Mauro's depiction of the Atlantic islands at the northwestern coast of Africa as well as the west coast of Africa down past Cape Bojador at the Western Sahara, which they passed in 1434.[4] Of course, that achievement of moving only slightly farther down the African coast, not even to the widest part of the prominent western bulge of the top of Africa, was still nautical progress at the time. Unfortunately, this Portuguese map copy is lost.

A second copy was commissioned between 1478 and 1480 by Lorenzo de' Medici and the Medici family of Florence but there was a new request attached to this copy. The Venetian aristocrat and intellectual Piero Dolfin (also Delfin or Delfino) had joined the Camaldolese order and lived on San Michele, and he was elected abbot in 1479. Dolfin had translated the texts on the map into Latin as an exercise, and, apparently, the Medici wanted their copy to have those Latin inscriptions instead of the ones in Veneziano which they could not read. We know these Latin translations once existed because, in 1494, the then-prior of San Michele, Bernadino Gadolo, wrote to Dolfin and asked for the originals so he could include them in a book about the history of the Order. Dolfin instructed Gadolo to look here and there, but those translations have disappeared. In any case, they were certainly used for the Medici map because, in the same letter, Dolfin writes that the copy was already sent to Florence.[5] No one knows what happened to the Medici copy of the Fra Mauro map. Girolamo Savonarola was causing havoc in the city and many bonfires destroyed "unacceptable" printed matter. The consensus is that Fra Mauro's Medici map was a victim of that censorship and that it presumably went up in flames.[6] But it might have influenced the Florentines. When in 1561 cartographers Egnazio Danti and Stefano Borsignori painted the fifty-three maps in the Hall of Geographic Maps in the Palazzo Vecchio in Florence for Cosimo I de' Medici, they clearly had seen, and studied, Fra Mauro's map.[7]

Although the Portuguese and Florentine copies have vanished, their commission by high-ranking, educated, and powerful people such as King Afonso V and the Medici family speak to the fact that this map was already renowned when it was completed, and it became something that others in power wanted to have. But it was two hundred years before Venetians remembered, in 1761, that they had this fabulous map hanging in a monastery on an island nearby. Just as the Venetian Republic was sliding into ruin, a group had been assigned the task of renovating the Sala Dello Scudo (Room of Shield) at the Palazzo Ducale (Doge's Palace) because room had been destroyed twice by fire, once in 1433 and again in 1483, and the original large maps on the walls were all in disrepair. The timing might seem off since the Republic of Venice was in a downslide and eventually ended thirty-six years later, but that was the point. The idea was to make a last gasp at promoting the power and might of the Venetian Republic by cartographically representing its domains and the rest of the world.[8] Historian Angelo Cattaneo says this renovation was a sort of "funeral epitaph" for the republic that had lost power and glory and that these politicians "were determined to remember it [the republic], even they couldn't save it."[9] This project was supposed to be a testament to the contributions made by Venetians to cartography, geography, and science.[10] They also held up precious Venetian globes as "instrument[s] vital to the transmission of a Venice-centered national historical memory."[11] To make their case that maps should be recreated for the walls, the committee turned to various Venetian explores such as Marco Polo, Niccolò de' Conte, Giovanni Caboto, and the Zen brothers, and they consulted and cited Fra Mauro's world map as an example of the contributions Venice had made to exploring and understanding the world. Of course, there was no way they could use Fra Mauro's map to fill in the missing pieces of the other damaged maps, because those were of a different time and construction.[12] Nor did they propose moving that map from the Biblioteca Nazionale Marciana into that room.

As the Republic of Venice surrendered to the French in 1797, and it suffered through various occupations by both France and Austria, Fra Mauro's map became of interest to another great power in the early 1800s. Instead of the map being used to prop up the honor of Venice, this time it was oddly appropriated to defend the march of Great Britain toward empire. Four British aristocratic men—the Dean of Westminster, the Earl of McCartney, the second Earl of Spencer, and the fourth Earl of Buckinghamshire—had a notion that they wanted Britain to possess a copy of Marco Polo's so-called original map, and they had been convinced

that Fra Mauro's map was an annotated version taken directly from
Marco Polo's travels. But no one knows if Polo made any maps, and if
he did, none of them survived his lifetime. They were certainly not in his
book which was, in any case, written by somebody else and long after
his journeys. These enthusiastic gentlemen also failed to realize that
Mauro's mappamundi was of the whole known world during the Middle
Ages, which meant it was very much more extensive than Marco Polo's
travels through Asia two hundred years before. Although Fra Mauro
had used the toponyms and various descriptions from Polo's texts for
the East, he had also relied on explorer Niccolò de' Conte and others.
But this group of Brits had been fooled into thinking Fra Mauro's map
was simply a copy of some maps by Marco Polo through the writings of
another Venetian, Giovanni Battista Ramusio. Ramusio had put together
three volumes of travel writing titled *Navigations and Voyages* (*Navi-
gazioni et viaggi*) that were printed in Venice in the mid-1500s.[13] The
first volume came out in 1550, the third in 1556, and that was followed,
out of sequence, by the second in 1559, after Ramusio was dead. It was a
grand project that aimed to gather and print the tales of various travelers
and explorers. Many of their journals had existed only in manuscript
form and so Ramusio was compiling them together as a three-volume
printed anthology. But Ramusio also made a major mistake. In the third
volume in which he included the writings of Marco Polo, he, as acting
editor, wrote an introduction to the section on Polo's travels and stated
that the great Fra Mauro Venetian world map was only a copy of some
marine chart combined with a world map and that both were of Chinese
origin and only annotated by the Polos. Ramusio calls Fra Mauro's map
"amongst the *various miracles* of this divine city," but he does not even
bother to mention Fra Mauro's name; instead, Fra Mauro is only referred
to as a lay brother of San Michele and dismissed as a simple copyist.[14]
Going down that rabbit hole even further, Ramusio didn't blame the
monk for any mistakes on the map. He claimed that Fra Mauro's map was,
in fact, a copy of a previous copy and that these mistakes were made before
the Camaldolese got involved. So, Fra Mauro was further demoted to

the position of being a copyist of a bad copy of Polo's map. As such, the British group felt Fra Mauro's map was as close as they could get to the great explorer Marco Polo if they wanted a cartographic display of the Venetian adventurer's travels.[15] And they did, because Great Britain was then colonizing much of the East. Reflecting that expansion, the people of Great Britain were going through an "Oriental" phase, which played out in fashion, design, and the arts, and that artistic turn underscored the righteousness of the expansion of the British Empire.[16] This twisted story with Ramusio at the center also suggests that Ramusio never went out to San Michele to see the map himself, because he would have noticed right away that it was a world map, not just the areas visited by the Polos. Perhaps Ramusio can be forgiven if he was simply trying to support his choice of including parts of the travels of Marco Polo in his travel volume. If he called Fra Mauro's map a copy of a disappeared Polo map, maybe those critics who had dismissed the famous merchant trader as a fabulist and a liar would be appeased.[17] In other words, Ramusio might have been a spin doctor for Polo, and that made Fra Mauro a causality of that promotion.

Based on Ramusio's description, the four British gentlemen, with funding from the East India Company, sent English painter William Frazer to San Michele in 1804 to make a copy of the map. Frazier had at hand the study of the map *Il Mappamundo di Fra Mauro* that had been published in Italian in 1806 by Camaldolese monk Placido Zurla.[18] That document was written at a time when monasteries in Venice were being repressed by the French occupation at the direction of Napoleon Bonaparte. In defense, Zurla wanted to show that the Camaldolese order had contributed a great deal to the corpus of human knowledge, especially science and culture, and so his book about the map was an attempt to stop the French from closing the monastery and its great library. It didn't work. The French occupiers were bent on wiping out, or drastically curtailing, the presence and power of religion in the city as they closed churches, convents, and monasteries and forever changed the very face and life of Venice. Even the great mappamundi couldn't stop the destruction and nothing could have saved San Michele as it was shut down and its irreplaceable, world-class collections dispersed in 1810.

In any case, in nine months, Fraser produced a copy of the map on vellum that was approximately the same size as the original map (7.8 x 7.5 ft or 239 x 229 cm) and it is a remarkably good copy, especially since it was, as far as we know, executed by just one person. Frazer also copied the legends line by line, retaining the original Veneziano, but he oddly exchanged Fra Mauro's depiction of paradise in the lower left of the map with the globe of the world climates on the right, without any explanation. Nor does the Frazer copy have the same gold chain circular frame that sets off the geography from the cosmology outside the earth as in the original, and it does not have the square gold frame surrounding Fra Mauro's whole creation. That Frazer map came to England in 1807 and now sits in the Reading Room of the British Library.[19] For many in Great Britain, this map underscored the idea that the British were heirs to what the Portuguese had started in the fifteenth century as they rounded the Cape of Good Hope and went east and expanded their empire into Asia. Fra Mauro's map was elevated to a piece of propaganda in service of an alien nation that would become one of the largest empires in the world.

Forty-three years later, in 1849, Manuel Francisco de Barros e Sousa de Mesquita de Macedo Leitã Carvalhosa—better known as the second Viscount of Santarém—and a minister of Portugal and a devoted historian, decided it was time to reproduce a copy of Fra Mauro's map. Viscount Santarém wanted to honor the copy of the map that he knew had been made for King Afonso V in the mid-1400s, the one that was lost. His purpose for making a copy of a copy (although he might have thought the King Afonso V copy was the original) was to extoll the history of Portuguese explorations during the Age of Discovery, particularly the Portuguese explorations down the west coast of Africa. It would be a "manuscript copy"—that is, a reproduction of the original in some other format or material. Instead of trying to make a copy of the original Fra Mauro map, Viscount Santarém took advantage of personal connections he had at the British Library and chose to work with Frazier's copy. The Santarém project took the map into a whole new realm of technique and possibility as he chose to use the lithographic printing process using etched plates. For this process, the lithographer draws the image onto a plate (or a stone) with a condensed grease pencil and then treats the open spaces with water, which of course does not mix with the grease. The water protects the open spaces and keeps the ink from smearing during the printing process. The surface is then treated with a chemical etch that bonds the grease lines to the plate. Once inked, a piece of paper is laid on top of the plate and the surface is burnished with a tool to stick the ink to the paper. Then, the plate is then covered with packing, such as a stack of paper, to equalize the pressure as it goes through a press. The result is a mirror image, which means the lithographer must draw in reverse. Doing that with Fra Mauro's map and all its intricate detail must have been a nightmare. Viscount Santarém hired a Parisian lithographic printer, who sectioned the map into six large sheets and placed them together to make the full map. As separate sheets, they could also be put into an atlas as folios, which he did between 1849 and 1855 for the Portuguese government.[20] This would be the first time Fra Mauro's map was printed rather than painted, and that atlas, with the copy of Fra Mauro's map, finally gave the map solid international exposure four hundred years after it was first drawn. Angelo Cattaneo suggests that this atlas was the first time there was an opportunity for a widespread and systematic study of an ancient map and that Viscount Santarém's execution and promotion of this printed copy of Fra Mauro's map initiated the study of the history of cartography, now an established academic discipline and an avocation for many who are captivated by maps.[21]

Twenty years later, the map was the subject of yet another kind of reproduction. In 1871 photographer Carlo Naya, an Italian photographer who was known for his fine photographs of Venice, made a life-sized photographic composition of Fra Mauro's map that was compiled from sixteen separate photographs, which he then hand colored. This photograph was one of the largest photographic presentations at that time, and it was certainly the only large-format picture of a map. Naya went on to win a medal for his map photograph at the 1873 Vienna World's Fair. Since photographs can be printed many times, and even giant ones can be put together to make an even more giant whole, Naya sold a black-and-white copy that was gifted to the Royal Geographic Society in 1873. The British Library and the Biblioteca Nazionale Marciana in Venice also have black-and-white versions, and there is an unpieced set of the black-and-white photographs at the National Library of Wales.[22] The value of these photographic prints of Fra Mauro's map is significant—there is currently an assembled and hand-colored copy for sale for approximately $300,000.[23]

Another seventy-five years on, there was another run of a printed copy of Fra Mauro's map. In 1956 the director of the Biblioteca Nazionale Marciana, Tullia Gaspartini Leporace, with the help of renowned Italian geographer Roberto Almagià, published a facsimile of the map on forty-eight color plates accompanied by the texts and a list of toponyms.[24]

Although the Fra Mauro map received all this attention spread over the centuries, it was not the star that might be expected for a map of its quality as well as scientific and anthropological value. At various points, it slid into oblivion, but some of that oblivion can be blamed on the Camaldolese order. The Camaldolese never elevated the skills of one monk over others, and with that philosophy in mind, it's possible that the cloistering of the map might have been on purpose.[25] There was also that problem of readability; the inscriptions were in Veneziano, not Latin, the language of the educated and aristocratic, and so intellectuals outside of Venice would not have been able to understand it anyway. The significance of the map, however, has always been understood by Venetians. Soon after his death, the government cast an honorific medal with Fra Mauro's image. This medal of tribute is a side view of an elderly monk in a monk's bonnet with the words *Frater Maurus S. Michaelis Moranensis de Venetii ordinis Camaldulensis chosmographus incomparabilis* (Fra Mauro, San Michele, Murano, of the Venetian Order Camaldolese, incomparable cosmographer). In the present-day this map has been renovated (again) and installed in the Museo Correr, the main art and history museum of Venice, in pride of place in a new exhibition of maps. It is still owned by the Biblioteca Nazionale Marciana, the main library of Venice, which means Fra Mauro's mappamundi belongs to all Venetians, as it should.[26]

CARTOGRAPHY AND THE
AGE OF EXPLORATION

By the fourteenth century, maps of all sorts had already been well integrated into most human cultures as means of communication, record-keeping, establishing ownership, and domination.[27] Maps were, of course, also necessary for travel directions overland and by sea as human groups continued to explore, and exploit, other lands. There were navigational portolans, regional maps, city maps, island maps, agricultural maps, property maps, and on and on. If something could be depicted in a drawing, it became a map. Mappamundi, the world map, was part of that cartographic explosion, but world maps were very different in comparison to, say, regional or city maps, or maps cataloging a narrow interest, such as agricultural products, or natural resources, which were all practical maps. Instead, world maps continued as storytelling, worldviews, and propaganda. But the accuracy of those maps improved during the Age of Exploration (which is traditionally dated around the early 1400s to the 1700s) as travelers and sailors explored new continents and brought home more geographic, oceanographic, and cross-cultural information than Polo, de' Conte, and other medieval travelers had seen before. In addition, there was the underlying motivation of "conquering" and "owning" these so-called new lands as Westerners baptized every place with Western names and ignored the fact that people were already living there.

Unbeknownst to European explorers, waves of ancient humans had, by two million years ago, walked out of Africa into Europe, across Asia, taken boats, or walked across the Bering Strait and down into North and South America.[28] They had also rowed or sailed from Southwest

Asia to inhabit islands of the Pacific and found their way to Australia
and onto Tasmania by thirty-five thousand years ago. Humans were
already inhabiting the tundra of the Far North, the jungles around the
equator, islands in oceans and seas, and traversing sandy deserts long
before Western explorers "discovered" anything. The story of the Age of
Exploration is not so much an amazing act of humans finally traversing
an empty globe as much as the story of the continuing trek of human
expansion that had been going on ever since our species stood up on
two legs and started walking across Africa five million years before.
Crossing the globe during the Age of Exploration was just comparatively
faster than all those other expansions into unfamiliar territory because it
happened in vehicles that moved much faster than walking or utilizing
oceangoing canoes. At their disposal were various modes of transport,
methods of navigation, nautical skills, and sometimes gifts to offer that
allowed them to invade other continents and countries, oblivious to the
endemic peoples who had lived there for eons. [29]

But still, Western explorers during the Age of Exploration deeply
influenced the overall knowledge base of the world and changed how
maps were made. For example, they repeatedly experienced people who
lived in many other ways; their descriptions of "others" were sometimes
born out of excitement and sometimes out of fear. The West also picked
up all that natural history now filling Western museums, all those dried
or preserved plants, seeds, taxidermized animals, animal skins, skeletons,
and rocks. At the same time, these travels pushed cartographers to make
more accurate geographic maps of the world. Eventually, the stories of
others and the natural history of a place dropped off world maps because
the primary goal was not so much to understand the globe as to exploit
it. In other words, mappamundi eventually morphed from encyclopedias
to straight cartography.

Geographers began to fill in all the blanks about the accurate size of
the globe and realize that there were more and different continents than
they thought. Long after Fra Mauro's map, geographers added North
America, South America, Australia, and Antarctica. They eventually

deleted their imagined landmass at the top of the world that was supposed to balance out Terra Incognita (which turned out to be Antarctica) at the base of the world when it became known that it was simply a sea of ice. Cartographers also had to accept that there was a very large ocean on the other side of the Americas that Westerners had never imagined. If all the voices of all the people who already inhabited all those places could have spoken, together they might have filled in those blanks for the clueless Europeans, but they couldn't, and so it took one aggressive culture motivated by profit—that is, European culture—to sail about and see with their own eyes how vast and full the earth is.

This process was also very busy, with many ships leaving many Western countries and sailing for months or years at a time, which meant, as time went by, geographic "discoveries" could be fact-checked against each other, which in turn made the information more reliable, more "scientific." For cartography, it was a feedback system. Expeditions came back, mapmakers integrated what they had found into new maps, and the science of cartography moved forward. The evolution of cartography was also deeply changed by the printing press. In Fra Mauro's time, maps were mostly hand-drawn, one by one. But by the late 1400s, mapmaking had already embraced techniques such as etching on metal or stone and the process of woodblock printing. With the invention of a printing press by Johannes Gutenberg in Germany around 1440 that utilized movable type both books and cartography gained a whole new way to make and distribute printed matter. That kind of press came to Venice in 1469, and Venice then became a center for book printing. Within that burgeoning business, especially spearheaded by world-famous printer Aldus Manutius of Aldine Press, the classics were reprinted in their original Greek because printers could use movable type with the appropriate diacritics in place.[30] Fra Mauro died just on the cusp of this revolution and it's impossible to suggest what he would have made of the possibility of making multiple copies of his map using this different technology.

And yet, embedded in this dramatic technological evolution, there was also a drag on world maps that, although Fra Mauro had tried to shake it, others continued to embrace. That heaviness was the influence of the Ptolemaic view of the world. Fra Mauro's map was well-known in the West and considered an achievement of geography, but it was also often overshadowed by the rediscovery and total acceptance of that 1,300-year-old Greek manuscript of Ptolemy (see chapter 2). Ptolemy's eight-volume *Geography*, a manuscript without any maps, had become the penultimate instruction book for drawing a world map after its translation from Greek to Latin in 1406. That would have been the translation available to Fra Mauro, and it left him unconvinced on so many levels. Of course, he had no Ptolemaic maps to consult, and no real reason to trust Ptolemy's projections for flattening out the Earth, but Fra Mauro knew of the hold this book had on geographers and intellectuals and how widely read and famous it was. So, the monk spent a lot of space defending his anti-Ptolemaic decisions. But what is even more interesting about Ptolemy is his tight grip on the history of Western cartography long after the Latin translation, even throughout the Age of Exploration.[31] There had also been another Latin translation of Ptolemy's *Geography* in 1467, eight years after Fra Mauro died, and with that translation came the first construction of maps using Ptolemy's written directions. If something in cartography can be called a "craze," the total acceptance of an ancient Greek manuscript with no original maps as the penultimate instruction for drawing a world map has to be considered a craze. Some historians feel the adherence to Ptolemy might have held back the development of cartography, while others feel it was the spark that pushed cartography forward, but no one denies its influence.[32] And that process can be followed in the history of world maps that came after Fra Mauro's initial spark away from medieval mapmaking.

WORLD MAPS AFTER FRA MAURO

Another world map, known as the Genoese map, appeared at about the same time as Fra Mauro's, in 1457, and unlike Fra Mauro's map, it's a typical medieval mappamundi and is in no way as progressive.[33] The Genoese map is a lozenge-shaped world depicted on parchment with the animal's neck to the west. Its oceans are a lovely blue, and the usual landmasses of Europe, Asia, and Africa are connected. But what marks this map as fun is all the sea monsters, mermaids, and mythical figures. It also has some inscriptions, but no one knows who drew it or why. It's called the Genovese map because there is a flag of the Italian city of Genoa in one corner, and like most medieval mappamundi, was presumably made as a decoration for a household. This map never achieved the fame or acknowledgment of Fra Mauro's map and so it makes sense to think of it as a commission for a prominent family's wall, perhaps for a Genovese merchant trader who would have appreciated its surprising depiction of one European ship in the Indian Sea. How that ship got there, since Africa on this map runs to the frame, makes it more of an aspiration than the story of a real Western trade ship that made it to the Spice Islands.

Two other world maps were also made in the last decade of the 1400s, both by Henricus Martellus Germanicus, a German cartographer working in Florence. Germanicus made them both blue and white with gold accents, and both have Ptolemaic projections. One is a bit more decorated than the other, but they both follow the general outlines of Fra Mauro's map by filling up the frame with land. Although drawn a bit later than the Venetian map, they have no hints of the Western Hemisphere, which was just being discovered. Following his lead, both have African tips that insert into the border but are circumnavigable.[34] Perhaps Germanicus had seen Fra Mauro's Medici copy in Florence before it was destroyed, but no one knows what might have influenced his choice to make his Africa a route to the Indian Ocean.

At the turn of the sixteenth century, world maps changed forever because Christopher Columbus had "discovered" America, although he denied it until his death and instead claimed he had stumbled upon some islands off the coast of China. Of course, humans had already walked or rowed to North America and then down into South America long ago, becoming the first Native Americans, and Vikings had already founded villages in Newfoundland, Canada, around 1,000 CE, but Columbus was the "first" European to put a foot in the "New World." That shocking discovery sent all cartographers back to their drawing tables to reconfigure the Atlantic and fantasize about what might lay to the west of those discoveries.

The first world map to include the new-to-Europeans lands was by the Castillian pilot and navigator Juan de la Cosa.[35] This map is drawn on a large piece of oxhide parchment in lovely colors (3 ft 1¾ in x 6 ft or 95 cm x 1.83 M) and it has a backstory. Juan de la Cosa, like Christopher Columbus, was no amateur explorer. He had been with Columbus and his fleet on two voyages as the owner and captain of the *Santa Maria*, one of the three boats of Columbus's fleet. Nor was he unfamiliar with the New World. After the two voyages with Columbus, he made three other trips west, including explorations along the coasts of South America with the conquistador Alonso de Ojeda, who named Venezuela after Venice, and the explorer Amerigo Vespucci. He participated in explorations of Panama, Columbia, Jamaica, and Hispaniola, and was one of the first Europeans to set foot on the South American continent. Dramatically, Cosa died in South America when he was killed by a poison arrow during a battle between conquering Spanish troops and indigenous Columbians.[36] While Columbus doggedly stuck to the notion that he had found a water route to China, and made his crews swear to that notion, Cosa drew something else on a map of the world. Those so-called Chinese islands, according to Cosa's cartography, were instead a landmass nowhere near China and they suggested the discovery of a completely new continent.[37] Not surprisingly, since Cosa was an experienced pilot and navigator, his map is not the usual mappamundi of the Middle Ages, nor is it a clear picture of the New World. Instead, it appears to be two separate scenarios. The map is oriented with west at the top, and presumably was meant to hang that way, but turned with west to the left where it should be, the map is 2 feet wide and a little more than 1 foot tall (61 x 30.5 cm). The eastern half is a portolan chart of the world with the typical rhumb lines crisscrossing the earth and it has the normal portolan style of place names hanging off coasts. The eastern section also contains the known continents of Europe, Asia, and Africa (which is clearly circumnavigable) executed with the rules of Ptolemy. The Eastern Hemisphere is also exactly in that fanciful style of older medieval maps with their various monsters, religious references, and the ever-present Gog and Magog, although Jerusalem is not at the center of those known continents.[38] But the most distinct, even shocking, part of the map is a splash of deep green that arcs around the western edge of the parchment that makes this map so significant. That huge verdant arc denotes a giant landmass sitting in the western part of the Atlantic Ocean and it is not shown with any sort of western connection to Asia, which it would if Cosa thought the Americas were China. For example, there is no green color on the eastern edge of Asia, suggesting a connection, and yet the far-left side of this area extends right to the edge of the map. The very use of green for this unexplored landmass sets it off as different from the eastern part of the map and seems to highlight it as wild land. Taken together, these cartographic decisions confirm that Cosa thought of Columbus's discovery as a new continent. Also, the bottom part of the greenness looks remarkably like a bent South America, and the top part, too, could be mistaken for the East Coast of North America falling over. The gulf formed by this continuously arching landmass might even be the Gulf of Mexico, and it is filled with islands, including Cuba and Puerto Rico, which Cosa must have seen with his own eyes. He even planted flags at the places in the Americas including where the Venetian Giovanni Caboto (John Cabot), while sailing for King Henry VII, landed in Newfoundland in 1497, the Spaniard Vicente Yáñez Pinzón explored parts of the South American coast and noted the mouth of the Amazon River in 1499, and Pedro Álvares Cabral landed at what is now known as Brazil and claimed it for Portugal in 1500, making this map extraordinarily up-to-date for a quickly expanding exploration of the New World.[39] Historians of cartography suggest the overall style is consistent with other maps made in Majorca at the time, a place where

mapmaking was big business at the turn of the fifteenth century.[40] And
this map is signed and dated. It was lost for a long time but discovered
in an antique shop in Paris and then acquired by Queen Isabella II of
Spain in 1853. It is now housed in the Naval Museum in Madrid.

At this point, Portugal was the center of exploration and Italians had
lost the title of best mapmakers, but map collectors were everywhere. In
1502 an Italian named Alberto Cantino was sent to Lisbon by Ecole I
d'Este, Duke of Ferrara, to buy the duke a map of the world that would
include all the latest discoveries. After all, the Portuguese had already
crawled down some of the west coast of Africa even in Fra Mauro's day and
they were now doing the same down the east coast of South America. Even
more significantly, the Portuguese explorer Bartolomeu Dias had rounded
the bottom tip of Arica in 1488 and Vasco da Gama had made it to India in
1499, also by rounding the tip of Africa.[41] The Duke of Ferrara wanted
in, he wanted a world map for his collection. So, he sent Cantino to Lisbon,
where he seems to have purchased a world map from someone. It says so,
in Italian, on the back, where the transaction by whom and for whom is
documented. References to this map often use the word *smuggled*, presumably because it was done sneakily, and the author of the maps is not noted.
Sometimes referred to as the *Cantino Planisphere*, this is an uncluttered sort
of outline map of the world, including the newest discoveries. It was drawn
with a new method based on latitudes and circles that calculated the rhumb
lines. And it looks nothing like the previous mappamundi since it has few
images and little text, religious or otherwise. Two cities on this map get
pride of place—Venice and Jerusalem—but the contrast with Fra Mauro's
visually extravagant map is startling, and perhaps it was a harbinger of
how culturally sparse modern geographical world maps would become.
The Cantino map was also lost for a time after it was sent from Ferrara
to Modena by a pope. But then the director of the Biblioteca Estense in
Modena discovered it in a butcher's shop, and it has been in that library
ever since. The Cantino map is important to the history of world maps
and is mentioned here because it updated Italians on the many Portuguese
discoveries, and their trade routes and an Italian copy became the basis

for the Waldseemüller map of 1507 (see photo insert).[42] It's impossible to know what Fra Mauro might have thought of the Portuguese mapmaking market, but he probably wouldn't have been surprised, given their many explorations, and the fact that he, too, checked out Portuguese maps because they influenced the cartography of his planisphere.

The early 1500s also had cartographers of world maps playing with all types of projections, trying to adjust to the addition of new continents and a new ocean. In 1506 the Venetian Giovanni Matteo Contarini designed a world map, which was then engraved on a copper plate and printed by Francesco Rosselli. Contarini used a "polar projection," which is Ptolemy's first suggested projection labeled as coniform, meaning shaped like a cone.[43] This map presents a God's-eye view from above the North Pole that fans out east and west. The result is a circular map missing the top two thirds. It is the first printed map to show the islands that Columbus ran into, and on this map, they are floating in the Atlantic and separated from both Asia and Africa.[44] The Contarini-Rosselli map was little known at the time and was lost for centuries, then found again in 1922. It now resides in the British Library. The Johannes Ruysch map world map of 1507–1508, made in what we now call the Netherlands, is also a cuneiform polar projection and shows the discoveries of both Columbus in the Caribbean and Caboto in Newfoundland. Interestingly, and incorrectly, Ruysch attaches North America to Asia on one side and Greenland on the other. This map was famous in its day; it was even included in two editions of Ptolemy's *Geography*, printed in Rome in 1507 and 1508, and had wide distribution. It has place names and mountains but little commentary about human action and is a good example of how world maps in the 1500s were becoming records solely of geography rather than compendiums of the beliefs that had covered medieval maps. Charmingly, this map does note the word *codfish* on the Grand Banks of Newfoundland, a nod to the growing importance of the fish staple that had spread all across Europe.[45] Perhaps Fra Mauro would have found these maps too sparse, too empty, and uninformative about the world, because humanity had been pretty well erased.

Another world map around this time that received scant attention for centuries, but now holds a special pride of place for Americans, is the Waldseemüller map. This world map is the first one to use the word *America*, and that word was placed in South America near present-day Argentina. The northern "island"—that is, present-day North America—is titled Parias. But it was the first map of the New World that used the word *America*, and that distinction is why the Library of Congress paid some German aristocrat $10 million for the only Waldseemüller map now in existence. This map is also known as "America's birth certificate."[46] What the Library of Congress has, however, is not one of the one thousand original prints but a reprint, but since all the others are missing, that was the only choice. This map was designed by Martin Waldseemüller and printed in 1507 by a group of scholars living and working in a small town in France using the woodcut method. There must have been an overall design on paper that was then divided up for woodblock printing; that's the only option for a map that was to be so large—7 feet (213.36 cm) wide and 4 feet 2 inches (127 cm) tall. With the woodblock process, a carver takes a block of softwood and with a knife cuts out the areas that are background or open spaces. Since these areas are recessed, they will not carry the ink to paper, but the ridges that are left hold ink and show. Like a lithographic print, this kind of printmaking produces a reverse process, a mirror image, and so the carver needs to take that into account, especially with maps that have complicated text and are meant to show accurate geographical spaces. The block is then inked and covered with a sheet of paper, often rubbed all over or burnished with a tool to help transfer the ink from the carved woodblock to the paper.[47] The whole thing is then topped with a flat heavy piece of wood or leather. Once printed, the twelve parts of this map could then be glued together as one world map.

The geography of the Waldseemüller map is based on the dictates of Ptolemy's second projection, but it also contains all the latest discoveries to that date, meaning long slices of land that represent North and South America with water beyond them, suggesting the Pacific Ocean which had not yet been seen by European explorers. The Spaniard Vasco Núñez de Balboa would be the first European to view the Pacific in 1513. Waldseemüller thought those lands we now call continents were big islands and not the East Indies nor China. The map also names Amerigo Vespucci in the title, and Vespucci and Columbus in an inscription in the top left that repeats the Roman poet Virgil's prediction that there was more land in the Southern Hemisphere yet to be discovered. Map historian Jerry Brotton points out that Vespucci, a Florentine navigator of some repute who had been on three voyages to the New World, in no way earned the right to have his name plastered across two continents. It is unclear if Vespucci even thought this "new" land, which he had also explored, was part of Asia or something else.

One of the greatest cartographers in the world, and a master of beautiful maps, Ottoman admiral Piri Reis, made a map of the world, actually a portolan chart, in 1513, but we only have the western third, which is housed at the Topkapi Palace in Istanbul. [48] That fragment does show the coast of Brazil, parts of Central America, Africa, and Europe, and it may have a hint of Antarctica. Reis also claimed in the text on the map that it was based on twenty sources, from Alexander the Great to Columbus, and there was excitement when it was rediscovered in 1929 because some thought it might be a copy of maps that Columbus was said to make during his voyages. Although Reis got many things wrong about the New World, especially North America, he was such a fine cartographer that he was able to place the South American coast at the right longitude relative to Africa, the first cartographer to do so. However, this map artistically continues the more medieval tradition of having monsters and weird beasts across the world.

Then came two events that forever changed the world of cartography— Balboa saw the Pacific Ocean in 1513 and Magellan and his crew completed their circumnavigation of the globe in 1522. Therefore, previous world maps were suddenly obsolete. Also, naval and trade powers started to divide up the world as they saw fit, and to their advantage, and maps reflected that process. For example, European mapmakers took hold of the world and centered their maps on the Atlantic, giving the impression, which continues today, that Europe, Africa, and North and South America are the centers of the world and Asia is on the backside of the earth. Cartographers also started to ignore much of Ptolemy because he hadn't known about the Americas or the Pacific Ocean, which made his projections worthless. World maps also became active agents of world change, not just decorative objects to hang on the wall. Historian John Noble Wilford states, "Exploration and cartography were now joined. No longer, after Columbus and Magellan, could mapmaking afford to be a contemplative occupation, the pursuit of cloistered minds."[49] For example, Giacomo Gastaldi, working in Venice, produced an atlas of maps in 1548. Although he followed Ptolemy and still had North America attached to Asia, Gastaldi's greatest contribution to cartography was his series of copper-engraved maps, which were pocket-sized and allowed a much wider audience to see the world.

Along with public access and commercial production,[50] world maps became radical political documents. According to cartographic historian Jerry Brotton, "Maps became legal documents binding rivals together in agreements based on geography."[51] For example, in 1529 the Portuguese navigator Diego Ribero made a map to underscore Spain's claim that they owned the Molucca Islands (now part of Indonesia), which were hotspots for spices. He even placed flags all over his map to denote which European countries owned what. His map, according to Jerry Brotton, was also a move toward the kind of scientific mapping we see today.[52]

The 1500s also produced one of the most famous of all cartographers, Gerardus Mercator, who solved the projection problem better than anyone before him.[53] In 1538 Mercator made a world map with an unusual polar cuneiform projection spread across two pages. On that map, North America and Asia are separate. It also has an Australia and a gigantic Antarctica. This map is the first time anyone used the names North America and South America to denote those continents.[54] It was also pretty accurate, although the edges of continents were faded to suggest the roundness of the globe. In 1569 Mercator made another world

map, but it was radically different from his previous attempt because he
had figured out a brand-new projection, one that changed everything.
This map is flat and covered with lines of latitude and longitude as
well as lines for the Tropics of Capricorn and Cancer and the equator.
The Eastern Hemisphere is the standard Europe and Africa, and the
Western Hemisphere shows large amorphous blobs of the North and
South American continents. In the southwest of that hemisphere is a
large island that he labels New Guinea. At the bottom of both hemispheres is a landmass we might assume is Australia, or an exaggerated
Antarctica, but is instead labeled Terra Australis—that is, unknown
southern land—a nod to the ancient Greek idea of a mythical land
down under. After all, cartographers reasoned back then, there must
be some land down there that balances out the Artic at the top of the
world. Wilford says the name Terra Australis comes from the ancient
Greek word *anartikos*, so it all fits together in a way and confirms that
ancient cartographers were convinced there was a large landmass at the
bottom of the globe even if no one had yet seen it.[55] Europeans also
had no contact with Australia until 1606, so Mercator had no idea it
was there. But no matter, because the significance of this map is not
the placement or size of the lands or oceans but how and why they
were projected into flatness. His point was to provide navigators with
the mathematics to help them draw straight lines across the map while
taking into account the curvature of a globe and being able to follow
a clear and consistent course toward a destination. There are, in fact,
navigational rhumb lines all over this map, making it much like a world
portolan chart. Mercator wanted those lines to be accurate and useful
for sailors, and he did this by lengthening the longitudes at the top and
bottom of the map. But in doing so he also distorted the continents.
Canada and South America are way out of proportion and too large
(South America is nine times bigger than Greenland, for example),[56]
but that didn't matter because this map was meant for navigation, not as
an object to hang on the wall. It's titled *New and Improved Descriptions
of the Lands of the World, Amended and Intended for the Use of Navigators.*

Although the calculations and lines on the Mercator projection were not perfectly accurate, pilots took hold of his mathematics, and by the next century they had figured out how to go long distances and end close to the mark. At the same time, the Mercator's world map of 1569 also harkens back to medieval mappamundi with its many monsters, giant whales, and Pygmy peoples at the North Pole. He also echoed Fra Mauro with trade boats bobbing about on the various seas and oceans. But this map was looking forward, and Mercator wanted it to be a resource for future navigators. The Mercator map shows North America and Asia as separate continents, and it divides the Eastern Hemisphere from the Western Hemisphere. According to John Noble Wilford, "By converting [world] maps from philosophical pictures and rude drawings to more useful tools, Mercator prepared the way for cartographers to enter the modern and more scientific era."[57]

Mercator's map essentially confirmed that the world was not just about discoveries, but about trade and commerce as in medieval maps such as Fra Mauro's, although some international maps were still focused on explaining the essential vastness of the world. Mercator's orientation and purpose, a reboot of Fra Mauro's aim to highlight trade shipping routes in the late Middle Ages, was later highlighted by Joan Blaeu of Amsterdam, the official cartographer for the Dutch East India Company. The Dutch East India Company was a commercial enterprise; they ruled trade from the East and made many of the citizens of the Netherlands very rich in the 1600s.[58] Blaeu produced a gigantic copperplate engraved world map in 1648 that was the size of Fra Mauro's map (6 ft 8½ in x 9 ft 10 in or 2.43 x 3 M) and that reflected the company's explorations in the East. Tellingly, this two-hemisphere map, like the Dutch East India Company, had little interest in the Americas at that time, and so his North America just fades away. At the same time, it also hinted at the continent of Australia, including Tasmania, both of which would have been within the reach, and of interest to, the Dutch East India Company.[59]

The Chinese had their first cartographic look at America in 1602 with the *Map of the Ten Thousand Countries of the East* (*Kunyu Wanguo Quantu*) by the Jesuit Italian priest and missionary Matteo Ricci. The toponyms and text were written in Chinese characters. Ricci spent time in Goa, India, and Macau, and was eventually allowed into mainland China and is said to be the first European to step foot into the Forbidden City in Beijing. He was allowed entry because he was a learned astronomer. Ricci died in China in his late fifties. He was deeply embedded in Chinese culture and philosophy. He learned Chinese and dressed in Chinese fashion, believing that assimilation was the key to converting people to Christianity. He also believed that Confucianism was not so far from Christianity. The *Map of the Ten Thousand Countries of the East* was his third attempt at a world map. The first was drawn on a wall and not in Chinese, but officials asked him to redo it in Chinese, which he did. Of course, those wall maps did not survive. Then Ricci made another with two Chinese collaborators, Li Zhizao and Zhang Wentao. This map is a woodblock print in brown ink on mulberry paper and there are six extant copies printed on rice paper. That method was well-known by the Chinese printers who also made screens that way; printing a map with six wood blocks was a simple process for them. This is also a very large map at almost 6 by 12 feet (1.83 x 3.7 M). The cartography is not particularly accurate, but for Ricci, the point was to demonstrate to the Chinese that a Christian God had made such a glorious world. The map is now sepia brown. What makes it so interesting is that it puts China, not Europe, at the center, which would have been a shock to Europeans. For the Chinese, it was the first cartographic look at North and South America, so this map was an exchange of cultures. Another Asian world map from the 1600s is called *Kunyu tushuo* and it echoes Ricci's map in the sense that its author was also a Western missionary, this time the Flemish Ferdinand Verbiest, drawn in 1674. Verbiest was also an astronomer and mathematician who spent decades in China. This map is two separate hemispheres with plenty of text in Chinese. One of two copies, it is now in the same exhibit in the Museo Correr as the world map by Hajji Ahmed and right next door to Fra Mauro's map. Significantly, the only other display in this room is a page from Marco Polo's will, a nod to the famous Venetian trader and explorer who was the basis for the Asian and Southeast Asian parts of Fra Mauro's map and of so many other maps. Another Asian world map is the Indian world map, which came much later in 1770, now in Berlin. It is much like the one Fra Mauro made two hundred years earlier—it is also a very big circle (a little more than 8 ft or 2.4 M in diameter)—and has all sorts of medieval creatures on the landmasses and in the water. It appears to be based on the stories of the fifteenth-century Islamic explorer Ibn Mājid, and the text is in Arabic, Persian, and Hindi. [60]

By the 1800s, all maps, including world maps, had blossomed into multiuse documents, and they were becoming international in their perspective and implementation. And yet, the early pattern of using world maps to overlay conquests, domination, and politics onto geography has continued into the present day. What Westerners are most used to is that standard map with countries delimited by different colors, with Europe in the center and all the oceans and seas a lovely blue. Although we think of this as *the* world map, it is, in truth, a political map. [61] We know this because the borders and colors of countries change all the time due to wars, treaties, and continued conquests, and so the world map continues to transform even though we now know quite well what the earth looks like.

At the same time, world cartographers also continue to focus on the issue of projection, always wrestling with how one could take the skin of a sphere and lay it flat and keep all the continents and oceans in perspective, a problem that can never really be solved. [62] But they continue to fiddle. In 1963 Arthur Robinson produced a world curved on the sides but flat on top and bottom, a projection familiar to all of us. It also has straight lines of latitude and curved lines of longitude that do not have the usual distortion of the Mercator projection. This map is a compromise of various other projections and is now considered the standard view of the earth laid flat. Then in 1973 the German historian Arno Peters moved the world map away from its usual focus on Europe at the center and proposed an "equal areas method"—that is, a cylindrical projection. [63]

The landmasses look nothing like the shape we are used to, but they are in correct proportion to each other. Architect and cartographer Abbass Bazeghi has also produced several world maps seeking the perfect projection. In 2016 he received a design award for a projection that split up with world into three lozenges, with the middle and right lozenges broken apart at their tops. This map is said to be at least 99.99 percent correct in its proportions.[64] It only took 450 years for all these corrections and still, the flat map of the world is not perfect.

In 2021 the *New York Times* asked readers the perpetual question, "Can this new map fix our distorted geographical views of the world?"[65] The article presented the latest attempt at projection by J. Richard Gott and a colleague at Princeton University. You can color print out two flat circular maps in polar projection, cut the circles out, and glue them together, one on top of the other, with the map side out, as if you were gluing together two vinyl records. That makes the world into a flat plate with the Northern Hemisphere on the top and the Southern Hemisphere on the bottom, but it's basically a squished flattened ball of the earth.

As such, it's also the truest flat projection of the world ever made, so far. But there will always be distortions in maps of the world, and in those distortions, we will continue to see the mind spaces of the various humans who make the maps as well as those who view and interpret them.

CHAPTER 8

Why World Maps Still Matter

*Long before the famous Earthrise photo taken by Apollo
astronauts, creating an image or map of the entire known
world was a singular human endeavor. Imagine a half a
millennium earlier when only those who had traveled could
describe distant lands. Imagine watching the hustle and bustle
of merchants and sailors returning to the port of Venice. What
stories could they tell to paint a portrait of our planet?*
　　　　　　　　　　—NASA, Landsat Science, NASA, 2014

*Maps are critical tools that can help deconstruct
violent events by providing a mental image of a
location and event in the onlooker's mind. They are
memorials for victims and their families, and they
provide evidence to bring about peace and justice.*
　　　　　　　　　　　　　—The Decolonial Atlas, 2022

No matter the age, world maps hold memories and also point to the future. Medieval maps were used as religious propaganda as well as instruments to prop up the identity of various countries or rulers. Today, just as in medieval times, cartographers continue to infuse world maps intended for public viewing with an array of political, social, cultural, and ecological issues, using geography as a template for making a global point. They can also be instruments of social change for the good, and this is why maps still matter in a very deep sense. At the same time, cartography has gone through extensive changes in format as well as in the technology used to make a map. These changes range from the invention of the printing press with movable type to the digitization revolution, super-computer analyses, and the availability of images of the earth taken from space. And yet, the goal of world mapping remains the same—to provide an up-to-date image of the world that can be viewed and instantly under-stood. The very nature of world maps is that they engage the eye as well as inform and educate. And they delight. Humans are, after all, visual animals; we need to see something rather than smell, hear, or feel it to absorb cognitively whatever it might be. Confronted with a photograph of the earth hanging in space or a world map made up of a patchwork quilt of countries, we can't look away. We are compelled to look more closely.

World maps, old and new, are bought and hung up and treasured by people who have no intention of traveling the world. The human obsession with world maps also includes the drive to make miniature replicas of home called globes. Long ago, in 150 BCE, Crates of Mallus constructed the first globe in Greece, and the oldest surviving globe was made in Germany in 1492 by Martin Behaim. We also have atlases to show the world from far away as well as close up. Globes, atlases, and world maps seem to matter even more these days as we go off-planet to other worlds, and as we see from space how our species is harming our home. To some, that view of Earth that the crew of Apollo 8 had in 1968 portends the future when humans might move to other planets because we want to or have to. Our obsession with a visual representation of our world comes from a deep-seated awe of where we live coupled with an unfilled longing to know every inch of this planet. Natural selection gave us big and complicated brains and we use so much of that mental power thinking about where we live, in a global sense.

World maps, in particular, are significant resources in all this knowledge-gathering because they chart our long-term interest in the world in an accessible form. Also, compared to globes, these flat maps, even when included in atlases, are more easily owned, read, and stored. World maps are also objects enlivened with nature as it evolves and the history of the various civilizations we have constructed. That, too, is why people still make world maps, because world maps are the human story symbolically rendered. Now they are also tools for research and change. World maps may have improved with cartographic discoveries, our perspective may have become more accurate with space exploration, and the techniques for making world maps may have exploded in the digital age, but the desire, the purpose, for making such maps remains the same—to know and understand our world as the object, the planet, upon which we live.

FINDING OUR WAY ACROSS THE WORLD
WITH DIGITAL TECHNOLOGY

The invention of computers and computer chips has changed the very nature of maps, which means, of course, that these new technologies have also dramatically changed worldwide human behavior relative to geography. The most significant development that radically changed the field of analytical geography was the invention of geographic information system (GIS) programs in the 1960s. With advances in computer database storage and analytical capabilities, researchers began to upload large data sets of information across the world and plot those complex bits of information across geographic space. In doing so, these scientists created a platform for analysis that could display visually the interaction of any number of variables and how they relate to geography. This approach is rooted in the history of choropleth maps, thematic maps that plot a subject across its geographic distribution. A French map from 1829, for example, mapped out education availability across France. Three years later, in 1829, Venetian geographer Adriano di Rodolfo Balbi and André-Michel Guerry plotted two variables—crime and education—across various French academies and royal courts. This map was significant for the development of GIS platforms because it showed that was something to learn, and something to be gained, by plotting an issue across geographic space.[1]

Today, GIS software systems combine digitized map images with digitized data, a marriage intended to be a research platform. When we read about some subject and its distribution across countries or cities, the hand of a GIS software platform has been at work. After all, this is the kind of work that computers are very good at—parsing variables and distributing them into categories across a distribution. Before computers, this kind of work was not just time-consuming, it was also prone to human mistakes and biases. But now, assuming the raw data have been added correctly, there is a universe of possible questions that can be addressed relative to place. For example, throughout the COVID-19 pandemic, the various world and local maps that appeared in the media are GIS products that take into account contagion rates, hospitalization rates and ICU beds occupied, vaccination status, and deaths distributed across country, state, and city. These maps often also plotted those variables according to age and race. This would have been a slow and massive undertaking before GIS and speedy computers, but now, those concerned can follow the numbers daily as others update the maps. The end result, for the viewer, is a real sense of individual risk at an individual place. More importantly, these data and their distribution are essential for epidemiologists, who track diseases as they emerge and figure out who might be at greatest risk. These mapping data have then affected the way people responded to the virus and evaluated various decisions such as vaccine boosters, let alone how individuals have run their social lives for the past few years. We have only been able to follow the directive to figure out our individual risk in any situation because we have been given these maps and data. In a sense, these maps are a cross of demography and geography, but not just for having a look. In this case, whatever the COVID-19 maps show on a particular day has had a demonstrable effect on what people do. They have been dynamic indicators of how the virus is behaving, what individual risk might be in a particular location, and they saved lives. The global and country maps, changing from yellow to red and back again, have also been explicit visual barometers of the survival of our species during a global pandemic.

But how do they make these maps?

First, researchers need very good pictures of the earth that they can translate into accurate maps. The geographic images of the earth used for GIS projects usually come from spacecraft, airplanes, and Landsat satellites. Landsat images are particularly useful for plotting data across the earth because they produce a continuous stream of digitized images taken in a one-hundred-mile-wide swath as the satellites circumnavigate Earth in polar orbit. That orbit also repeats itself every sixteen days, providing a dynamic record of changes on the earth's surface. NASA, in collaboration with the US Geologic Service, maintains this record of images. The stated goal is "to chart environmental change" on Earth, and they've been doing that since 1972. With continuous digitized images of Earth, Landsat images have tracked the melting of glaciers, deforestation of jungles, the encroachment of human populations, and changing land use throughout the world, among everything else that changes on the earth's surface. GIS scientists also use images taken from other satellites and airplanes, but Landsat is the best and most accurate source for images of Earth taken from space, and they are available to everyone.[2]

Since Landsat images are readily available to anyone, it's also possible to take those images, focus on them according to some question, layer other digitized information on top, and produce a visual composite. John Nobel Wilford calls this process "cybercartography," because everything is digitized and held in cyberspace.[3] He also describes this process as linking geography, "where things are," to descriptive information, "what things are like."[4] GIS projects are so interesting and useful because they are based on masses of data that can be crunched by supercomputers, easily connected with up-to-date images of the earth's surface, and viewed by others far beyond the researchers' home group. After all, the multilayered concoctions can also be uploaded to the cloud and shared.

GIS software is also a very useful tool because it can be played with—that is, manipulated in informative ways. Once the full picture is assembled, various clicks can change the variables, or even fool with the geography as in moving rivers, or, say, having water inundate coastlines.

With the ability to create changes in the variables, including the geography, this integrated, flexible map then becomes an experimental, seemingly alive, map that surpasses its mappamundi forbears. Unlike those older maps, GIPS can be predictive as the researchers test various options and manipulate the data points to see what might happen; GIS goes beyond just knowing where things are to the role of testing how things might be if humans choose to do this or that with the earth. GIS project maps have sprouted up in government, for example, and they are now an accepted methodology for the population census, in military decisions, to help plan for disaster relief, and city planning. Since its introduction, GIS software has also become readily available to the amateur geographer, and so who knows what might be mapped next or how that information might be used by people, communities, and countries.

While many uses of GIS software are local, or by country, there is also a global option when considering an idea or a problem, because this system of analysis can cover the whole planet. As a result, GIS is a potential tool for understanding and confronting the ravages of climate change, diseases, and famine. It has also been used to pinpoint clean water sources and has helped in predictive ways during weather-related disasters, such as hurricanes and tornadoes. At its most fundamental, GIS can keep track of the increase or decrease of human populations throughout the world. In other words, this high-level global, integrated, multilevel analytical geographic tool has the potential to address, and provide ideas on, universal humanitarian needs. It is the present-day mappamundi, but unlike its predecessors, GIS maps have the potential to influence everyone on Earth based on geography.

Surprisingly, the Landsat group upon which much GIS work is based recognizes the importance of Fra Mauro's 1459 map and its place in the canon of visual representations of the earth. His map has its own Landsat page maintained by Landsat Science and NASA.[5] There you see the medieval Venetian map placed in juxtaposition to an image of Earth floating in space. This is not like the Anders photograph from 1968, but a composite and highly accurate *Blue Marble* image constructed by NASA from satellite photos. It has also been manipulated—that is, moved about and cropped with clicks on a computer—so that the *Blue Marble* image lines up in the same orientation as Fra Mauro's ancient point of view. They are both upside down and focused on Africa, Europe, and Asia. Set side by side, the two images—one drawn in a monastery in the fifteenth century, one constructed in a visual imaging lab and based on photographs taken from outer space—have the same general outlines of continents and coasts and bodies of water. According to NASA, Landsat Science: "The comparison is stunning when you consider that Fra Mauro compiled his data from the travel tales of myriad fifteenth-century sailors."[6] Fra Mauro, who never saw anything beyond his native city, and certainly had no idea what the world looked like from above, got so much right. Here, we see past geographic information tested with images from the present, confirming that the past informs the present and vice versa. Both images are human productions of how we see the world and there is a kind of constancy, a sort of faith in the process of human imagination, when seeing that an image from the 1400s still holds up, even in the face of expanding knowledge and improved technology.

Other technological advances in geography have also changed human behavior, but this time the effect has also resonated daily on an even more personal scale. The invention of the Global Positioning System (GPS) has made navigation more intimate than reading a map and so much easier. GPS has a military history. It was first invented in the 1960s when US satellites tracked their nuclear submarines based on radio signals gathered by satellites.[7] By 1978, the US Department of Defense launched those sorts of satellites into permanent orbit to create a navigational system based on those same radio signals. It took until 1993 for the system to have twenty-four satellites orbiting the earth with the ability to pinpoint places across the globe with incredible accuracy. Although that system is managed by the US Air Force, and it is there for military, civic, commercial, and scientific needs, just about everyone uses it today on their phones and in their cars rather than paper maps. Although it feels like we all own GPS, there are two levels of this service—one for the public, which we, meaning anyone in the world with a device capable of interpreting the signs, can access for free. That level is called the Standard Positioning Service (SPS), and it's the one that has been commercialized with pins for restaurants, lodgings, entertainment venues, and stores. But there is another, more precise level called the Precise Positioning Service (PPS), which is only available to the military and the government. Think of it this way, driving from Philadelphia to New York City using GPS on my phone or car, I receive two options I can take, one a few minutes faster, and am told there will be tolls. That's quite a different sort of trip mapping than a drone strike on a military target in a country halfway around the world. Still, the convenience of using GPS for things like finding a restaurant or the way across town is based on the acceptance that this convenient global mapping technology was developed for defense purposes. In some ways, that tacit acceptance of the underlying reason for GPS is much like a medieval monk knowing his map could be used for military purposes as well. As to the commercialization of GPS, both Fra Mauro's mappamundi and the GPS system were designed to help that sector and encourage global economics. In any case, maps, no matter when they are made or who they are made by, usually have layers of purpose, and some of those objectives might feel uncomfortable to some.

There are, of course, many who still have a paper map at hand while listening to GPS directions because sometimes the instructions are wrong, the system can't distinguish between two places with the same name, or the directions come too quickly. Also, there might be spots where there is no reception at all, which means we get lost because there are no directions to access digitally. For example, using GPS in Venice, with its labyrinth of streets, and where many streets in different parts of town have the same name, is a dicey business. Everyone but Venetians always gets lost at some point, and GPS is not much help. One should always also carry a map of the city. This brand of traveler, the ones who carry a printed map with them (and I include myself in this group) harkens back to the idea of humans as visual animals who like to see where they are going rather than trust someone else giving directions verbally.

For me, there is comfort in seeing a map and noting the wider view of where I am going. I can then mentally note the endpoint and orient the map in that direction, choose among various routes, and then set off. The whole trip is then etched in my neural networks and off I go. Although it's an old fashioned option these days, I also like a backup paper map at home in the United States in case the GPS doesn't work or if I travel outside cell phone reception.

In any case, we now pretty much have every place on Earth positioned exactly with points of coordination that are provided by radio waves bouncing off the earth and going back to satellites that collect the data. A section from Google Maps is simply taking those GPS coordinates and building a static map from the information. Every time we look at a map on the computer or on a smartphone, we are relying on those satellites orbiting the earth that are constantly giving back geographic pinpointed data. You could, of course, read these data with a dedicated GPS device, but now we all have it at our fingertips on our smartphones, computers, tablets, and even smartwatches. Google Maps and Apple Maps also allow us to see any place filled in with restaurants and shops and because GIS is incorporated into those maps, so we are visually shown how to get to those places.

On a more spectacular scale, digitized Google Earth is the most modern mappamundi of all. Launched in 2005 and referred to as a "geo-browser," or "geographic browser," it incorporates every possible digitized geographic resource available to patch together a visual representation of every corner of the earth, including under the oceans. The result is a three-dimensional ball or globe and not a flat projection. That globe can be endlessly tilted, rotated, zoomed in and out, and viewed in 3-D.[8] Google Earth is meant as a purely visual experience because it is not able to make the kind of analyses that GIS software can do. But it can be so much fun. For example, you can pop up political boundaries—that is, the boundaries of countries—or leave the earth as its natural self. Google Earth also has a "street view" as part of its package as you zoom in and switch to Google Maps, which means you can walk virtually through a foreign city or down a stream bed with the click of a mouse or the tap of a finger. Or go 3-D and see what the buildings and forests look like. Once I took myself on a flyover across the Venetian Lagoon to see what it looked like at the edges, estuaries, swamps, and shoals that lie beyond the typical tourist experience. It felt like I was flying low across the lagoon in a helicopter, a helicopter that could not just bank left and right but one that could also hover for as long as I wanted over a tiny uninhabited island where I could then see in 3-D the crumbling ruins of a palazzo or convent left long ago. Google also now has Google Mars, Google Moon, and Google Sky, which takes us, digitally, into space and onto other planets. These images are not made by humans walking in space—they come from the hands and minds of humans who built the satellites and wrote these grand programs to bring an accurate geographic picture of the earth and sky to our digital devices.

Google Earth is also trying to do some good. I recently took a Google Earth trip through a feature called Sea Rise Levels and the Fate of Coastal Cities.[9] I am interested in this topic because of the recurring bouts of *acqua alta* in Venice and the damage that water causes to the city. The eight-page presentation by Google Earth shows in images and text that 760 million people worldwide are now in danger because of coastal flooding. On the second page, the earth spun around, and then zeroed in on London in 3-D. The text explained London's industrial history and the effect of burning coal on the environment. The next pages spun to New York City, with a focus on Wall Street, which is surrounded by water; Mumbai, on the west of the Indian Peninsula where eleven million people are at risk; Shanghai; Lagos; Rio de Janeiro; and San Francisco. Following the adage that a picture is worth a thousand words, and following in the footsteps of medieval cartographers, Google Earth brings its point home not just by talking but by showing; subjects chosen by the Google Earth team are effective because they are presented visually as well as with text. The Time Lapse feature on Google Earth is especially scary because it pulls up a dizzying stream of images taken by satellites over thirty-seven years and shows what humans are doing to their planet.

For example, we can watch the Amazon Forest being turned into soybean and palm oil farms and cattle ranches. Along the way, this spinning Google Earth also takes you to other cultures and other places that are working against deforestation.

The loveliest tab of Google Earth is Fragile Beauty, which flies across the earth to some of our planet's most memorable places and discusses how the lands and seas are changing at all these sites. Feeling low? Click on the tab that will take you on a tour of cherry blossoms blooming all over the world, check out some possible rock-climbing destinations, or learn about the women working in conservation. Google Earth is an armchair adventurer's paradise and a place where anyone can have a wider, more global view of both nature and humanity. It is, in our time, the best mappamundi ever. Sure, it has faults and biases, but what cartography doesn't? And with this one you get to ride the globe as it spins and flies down to Earth to land on a place, a city, a group, or a change.

If we didn't know better, we might call Google Earth magic, not science. And like world maps of long ago, this spinning GIS creation is also a knowledge aggregator in the best sense of the word. [10]

THE USE OF DIGITIZED MAPS TO
UNDERSTAND THE PAST

While in Venice for the fall of 2019, I saw an announcement in the local paper, *Il Gazzettino*, about a seminar on the Fra Mauro map. I was just thinking about writing this book, and here was a chance for some academics to tell me why it was worthy of study. The presentation was held in the Sansovino Reading Room attached to the Museo Correr, which, in turn, is connected to the Biblioteca Nazionale Marciana. The researchers came from Ca' Foscari (the University of Venice) and they had been collaborating with a group from Singapore and some Chinese exchange students in Venice. The PowerPoint set up in the front of the room showed Fra Mauro's map with a rash of dots across its surface. The researchers had digitized both Fra Mauro's map and Marco Polo's travels to explore long-ago Venice and they were addressing all sorts of hypotheses about both historical documents.[11] How correct was Marco Polo in his descriptions? How accurate were the geography and the toponyms of Fra Mauro's medieval map? With the Chinese students, they were also able to confirm many place names in the East that Fra Mauro had taken from Marco Polo's travels, but no one had understood these names until the students had decent visual access to the writing on the map, which could be much more easily done and shared after it was digitized. The group found that Fra Mauro pretty much got everything right, which meant so did Marco Polo. This presentation was by an international group that wants to use modern technologies to understand the past by digitizing old documents, maps, and sites, and taking a meta-based worldview that brings together history and the present. Their project is called Engineering Historical Memory, and their products, so far, are many academic articles[12] and a whiz-bang website that is fun to play with.[13]

On this interactive site, the Fra Mauro map can be viewed as normal, or with an infrared, ultraviolet, or tangential spectrum. The ultraviolet spectrum makes the texts pop out. The tangential and ultraviolet views on the website highlight that the map made of velum was put together from four long strips.[14] But the most important contribution of Engineering Historical Memory through Fra Mauro's map is that it provides an overwhelming, thorough, and entertaining digitized addition to the surface of the map in the form of references for just about every spot, be they place names, geographical features, or text. Click on any point on the map and you are taken to a small Wikipedia box that provides the number of publications, images, and videos about that location. Click on that and the various specific resources show up. For example, a click on the point of the region of Goçam, or Kingdom of Gojjam, in present-day Ethiopia, and see that there are 193 publications, 200 images, and 1,234 videos to check out. Clicking on each one of those possibilities and see the lists of academic publications about the region, images, and videos. It takes you, in one instance, to articles such as "The Archaeology and Economic History of English Clay Tobacco Pipes," published in the *Journal of the British Archeological Association* and can be quickly accessed with another click. Noted images of a region today are accessible with a click and the available videos are a panoply of people and places from that region. Some of the drawn images on their version of the Fra Mauro map are also animated and they can be rotated 360 degrees so that we get a full understanding of items such as ships as they rotate after being virtually lifted off the map and their other sides digitally mirrored and incorporated. [15]

This interactive platform is Fra Mauro's map on steroids, the information brought up to date and bought alive with photographs and videos and given deep substance with academic articles. The result is the ultimate marriage of a medieval map as a knowledge aggregator brought into the twenty-first century with all its technical advantages. In this case, digitizing Fra Mauro's map and comparing it with Google Maps and adding the Wikipedia boxes with current references is a perfect example of modern technology brought to bear on a historical map. *The Travels of Marco Polo* was also digitized by this group and superimposed on Google Maps, providing an interesting visualization of his travels. Also, once superimposed on modern maps, many of the heretofore incomprehensible Chinese place names phonetically spelled by Marco Polo and copied by Fra Mauro suddenly made sense because the Chinese exchange students could trace them to modern cities. According to the presentation, the researchers found that, in general, Fra Mauro's map was surprisingly accurate.

The Engineering Historical Memory researchers aim to bring history to life with new technologies and new ways to research other ancient works and ancient places. [16] They have also digitized the Genoese map of 1457 (see chapter 7), the travels of Ibn Battuta in the 1300s, *The Morosini Codex* of 1500 Venice, and many other documents. [17] They have also played with the geography and mapping of an ancient Greek fortress in Messenia and a series of six hill forts in Rajasthan. The world and the history of the world are their oysters, and it will be exciting to track where they go with this technology as they dive into the past.

MODERN WORD MAPS AS CHANGEMAKERS

In 1982 Buckminster Fuller designed the *Spaceship Earth Map*, an icosa-hedron world map with twenty faces that could be folded and unfolded, and it was intended to make a point. This world map has no country boundaries, no up nor down. Fuller wanted to show a world without the usual political boundaries, those world maps we are so used to with their differently colored countries that divide up the world into parcels, and ones that are often in dispute.[18] While mappamundi certainly had their points to make, Fuller's map, and others like it, are trying to make the Earth more a collective, an alternate sort of world map that highlights the universality of the human species and the possibility of a world that need not be sectioned into potentially opposing parcels called countries. Fuller's map is geographically correct and yet startling because we have become so used to world maps that highlight our tribalism rather than our unity.

Others using world maps to focus on what is currently happening to our species on this planet have moved from one world map to many, by utilizing the atlas format—that is, as a book rather than a map to hang on the wall. But the point is the same—to visualize various human actions and their consequences around the world. The best known is *The State of the World Atlas*, which hit its tenth edition in 2021.[19] The original authors of the first *The State of the World Atlas* were the cartographer Michael Kidron in addition to Ronald Segal and Dan Smith, both activists for social justice and peace. As a result, this atlas is no mere piece of geography. Their thesis is to show what the real state of the earth is in terms of human health, migration, available resources, such as water and food, literacy, gender equality, and even COVID-19, among many global contemporary subjects that affect all of us. It is a wake-up call to the condition of the human species across the globe. A map that tracks human crises also takes constant attention beyond published books and atlases. A British group called Worldmapper highlights social and environmental issues and makes their points about inequality by shrinking or ballooning countries and continents accordingly.[20]

That same kind of cartographic political action is demonstrated by a website called the Decolonial Atlas, which fights for Indigenous rights and social justice by exposing injustice on maps. They state, "The Decolonial Atlas is a growing collection of maps which, in some way, help us to challenge our relationships with the land, people, and state. It is based on the premise that cartography is not as objective as we're made to believe."[21] Furthermore, as the Decolonial Atlas points out, the orientation of a map, how it deals with projection, if it has political borders, and whatever the mapmaker chooses to add or omit is subject to bias. This project was started by a group of Native Americans and many of their maps deal with issues of Native American languages, peoples, and cultures, as well as the economics of Native American life. For example, they have developed maps with original Indigenous toponyms rather than colonial designations, shown routes traditionally used by Native Americans that have been overlaid by current highways, and plotted the graves of Indigenous children who were forced to attend the residential schools in Canada. But they have also branched out into all sorts of directions and for all sorts of issues that also fit into their "decolonial" moniker. There are maps displaying killings by the police according to race in various large US cities, the cost of car insurance versus health insurance across the United States, and who voted to reject the 2020 presidential election result in the Senate and House of Representatives by state and political affiliation. They also provide information for those in need, such as *Bus Routes to Abortion Providers in the South*, as well as calls to action, such as the map that shows where to join blockades of land grabs. Their reach is also worldwide with *Who's Responsible for Climate Refugees*, and *Plastic Waste and the Top 25 Companies Producing It*. Some maps are seemingly nonpolitical, such as *Street Trees of NYC, London, and Melbourne and Where they Originated*, but even there, it's the plight of city trees that is of issue. They make a case for planting native species that would do better than imported species. Spending time on the Decolonial Atlas site is a visual learning experience and can be life-changing because their maps usually address hot-button issues and ones that most of us know nothing about or don't fully understand. Because their maps are thematic and based on data that can include numerical values and precise place names, they are often both historical and current, one overlying the other. Most importantly, the maps of the Decolonial Atlas highlight change over time, underscoring the fact that what we think we know is always imbued with hidden history. By taking that approach, they sometimes emphasize negative predictions for the future by showing the present as well as the past. But at the same time, the visual impact of these maps comes with a kind of hope that once viewed, there will be a call to action. In that sense, the Decolonial Atlas is working with the visual rather than textual axis for education and persuasion. This organization is using maps not only to educate but to change the world, and in that sense, collectively the Decolonial Atlas is also a mappa-mundi for our troubled modern times.

FRA MAURO ON THE MOON

One of the undebatable facts about the Fra Mauro map is its enduring presence, right up into the present. We see that in the website Engineering Historical Memory, in the myriad references to this map online over many platforms, the fact that it's easy to buy a reproduction of this map as a poster or puzzle, and in its current place, newly renovated and set as the showstopper in a large map exhibit in the Biblioteca Nazionale Marciana in Venice. But there is another tribute to this map and its creator that is otherworldly in its execution.

In 1935 the International Astronomical Union approved naming a fity-mile-wide crater on the near side of the moon after Fra Mauro. The astronomers who wanted that name on the moon knew about the map and its significance, and they honored a fellow geographer even though his work was of the earth. [22]

The name Fra Mauro also became more of a household name in the United States when, on February 5, 1971, astronauts from the crew of Apollo 14 landed in the area north of the Fra Mauro crater, a highland region known as the Fra Mauro formation. Apollo 13 was supposed to land there, but that crew famously ran into trouble and had to return to Earth without landing on the moon.[23] Members of Apollo 14, which included Alan Shepard, Stuart Roosa, and Edgar Mitchell, were the third group of astronauts to land on the moon, but this time their target was the lunar highlands, a more geologically interesting area than the ones visited by previous crews. After traveling to the moon for four days, Shepard and Mitchell used the lunar module *Antares* to leave the mother ship *Kitty Hawk* under the care of Roosa and they spun off and landed in the Fra Mauro formation. They went for two walks during their stay, pulling a rolling cart full of equipment with them. Humorously, there is an iconic photograph of Mitchell examining a paper map of the moon as they find their way. The purpose of the walks was to make scientific tests and to gather moon rocks and bring them back to Earth.[24] Then there is the iconic image of Alan Shepard hitting a few golf balls across that barren landscape with a combination of a golf club and a piece of lunar equipment. This mission to the Fra Mauro formation also had a small project that brought a hint of the moon back to Earth. Inside the main capsule, Roosa had brought along a sack of two thousand tree seeds provided by the US Forest Service. Once the astronauts returned to Earth, those "moon tree" seeds, from five different tree species, were planted all over the country.[25] And so, seeds that went to the moon have grown into trees in various parks on Earth; there might be one in your town.

The history of humanity as one species, one people was dramatically demonstrated when those astronauts went to the moon, took a few strolls across the Fra Mauro formation, and telegraphed images back to Earth from a lunar spot named for a Venetian cartographer who also changed the way humans viewed their planet.

❊

We are the cartographic species. We have drawn and photographed every bit of this planet, compelled to make sense of our world and our lives, to understand our place in the universe. This compelling urge has also led us, inexorably, to mapmaking. Fra Mauro embraced that urge in his time and in doing so he handed us an early road map of our place in nature. At the other end of time, Alan Shepard and the crew of Apollo 14 acknowledged the great contribution of that medieval cartographer when they stepped onto the moon at a spot named for him. Tellingly, Shepard's first words as he placed his foot on the Fra Mauro formation were, "And it's been a long way, but we're here." If Fra Mauro was listening from his place in the cosmos, he surely nodded and understood, because the human drive for exploration and understanding of who we are and where we live isn't over. It has just begun.

Acknowledgments

To quote Fra Mauro, "Furthermore I say that in my time I have tried to validate written sources with experience, researching for many years and profiting from the experience of trustworthy persons who have seen with their own eyes all I have faithfully put forward here." In the complicated history of cartography, and mappamundi in particular, I have tried to adhere to common approaches and theories and to substantiate what is known and unknown about this map. The map itself is a cypher, which makes it seductive and fascinating, but also a difficult subject. For that reason, I am deeply indebted to scholars of the Fra Mauro map Piero Falchetta and Andrea Cattaneo, whose groundbreaking and thoroughly researched books made this book possible. I only hope I have done them, and their work, justice.

I am also so very thankful to the online library system of Cornell University and the University of Pennsylvania, without whom I could not have written a word. It was not the same, being locked out of university libraries for two years, not being able to crawl through the stacks, but with these online systems, I managed.

I am also thankful to my literary agent, Wendy Levinson of the Harvey Klinger Agency, who saw the potential for this book right from the beginning. I had always wanted to write a book about one item that changed the world, and this was my chance. Wendy completely understood why this map captured my imagination and she trusted me to get the Venetian history and the history of cartography correct. With this and my other book on Venice, *Inventing the World: Venice and the*

Transformation of Western Culture (Pegasus Books, 2020) she has been a stalwart fan, trusted editor, and efficient advocate for my work, always going far and above what is normally required of an agent.

I am also grateful to Pegasus Books for their usual lovely covers, great book design, and enthusiasm for my writing and for this topic.

On a personal level, I am especially thankful to many friends who have repeatedly asked about the progress on this book, especially during the hard times of the COVID-19 pandemic, which left me inside my apartment, alone, with only my computer and my cat for two and a half years. I am not sure how I would have gotten through those days without the many phone calls, texts, emails, and Zoom calls for company. Thanks to Deb Elkington, Nancy Zahler, Judy Berringer, Becky Rolfs, Andrea Fox, Juliebeth Corwin, Laura Nelkin, Tamara Loomis, Steve Cariddi, and Sarah Douglas for being there.

Sandra Quinn, to whom this book is dedicated, pretty much never left my side during these hard times. She was always there, at the end of the phone line with an encouraging word, an interesting discussion, and questions about how I was doing. Sandy, an artist, also helped me figure out some of the sticking points concerning medieval artistic technologies that I wanted to describe about the production of this map. But it's the fifty years of friendship, the ever-present support, the good advice, and the fact that she always laughs at my jokes that earned this dedication. I also know that we will soon be standing in front of this map and looking at it together because Sandy's that kind of friend, one who really wants to see what I am doing and share with me the enthusiasm for a subject.

I also need to single out Steve Cariddi, a trusted friend who edited parts of the manuscript. Steve has also offered to show me the Fra Mauro formation on the moon, and that in itself rates a special thank you. Next time we are together, we will take his giant (to me) telescope and head for a hilltop in the middle of the night. Then Steve will show me the glory of the night sky as only a stargazer can, and I'll see for myself where Apollo 14 landed, and that event will wrap up this book for me in a nice, neat package. *Non vedo l'ora.*

Stefano Chiaromonte has been my Italian language teacher for over five years. During the pandemic years, we met on Zoom twice a week, and that schedule certainly kept me grounded, even sane. We did *conversazione* and *grammatica* every week, no matter the COVID-19 numbers, no matter my fears. In truth, I spoke to Stefano more than anyone else during that time. And all those lessons allowed me to read many articles in Italian that would be inaccessible to me without his lessons. When we finally met in Venice in April 2022, it was a meeting between friends, and we had a great time looking at the map and rowing up the Grand Canal. You can see the video of his visit to Venice on his YouTube channel, Teacher Stefano, and enquire about Italian lessons at www.teacherstefano.com (which I highly recommend!).

My daughter, Francesca D. Merrick, first sent me on this path of writing books about Venice. She is the person I trust the most in the world, and her opinion about my various ideas about book projects are always intellectually informed, editorially spot-on, and infused with her special understanding of who I am and what I can accomplish. The encouragement she gives me is vital to my life as a researcher and writer. Francesca also puts up with my many visits to Venice (although not so much during the pandemic) when we lose precious time together. I only hope the pastries from Pasticceria Tonolo, the Kinder Surprise eggs, and the very many pairs of Venetian bead earring make up for my absences. As always, and forever, *Mi so tua e ti, ti xe mia.*

—Meredith F. Small
Philadelphia

Notes on Illustrations

Unless stated otherwise, all photographs are in the public domain and covered by U.S. copyright

Figure 1: Fra Mauro's map, housed at the Biblioteca Nazionale Marciana, Venice and currently exhibited at Museo Correr, Venice. Photograph by Meredith F. Small, used with permission.

Figures 2–4: Fra Mauro map details. All images in the public domain.

Figures 5–11: Photographs by Meredith F. Small, used with permission.

Figure 12: Mosaic of Marco Polo, Municipal Palace of Genoa, Palazzo Grimaldi Doria-Tursi. Public domain.

Figure 13: From *Isolario*, Venice, held at The Morgan Library & Museum, New York City. Image in the public domain.

Figure 14: National Aeronautics and Space Administration (NASA).

Figure 15: The Babylonian world map is on display at the British Museum, London, U.K. exhibit number (G55/dc16). Image by Osama Shukir Muhgammed Amin used with permission of Attribution-Share Alike 4.0 International license https://creativecommons.org/licenses/by-sa/4.0/deed.en

Figure 16: Reconstructed in 1879 by Edward Bunbury for his book *A History of Ancient Geography by the Greeks and Romans form the Earliest Ages till the Fall of the Roman Empire*, John Murray, London. Image in the public domain.

Figure 17: Image displayed on https://i1.wp.com/muslimheritage.com/sites/default/files/map2_0.jpg Image in the public domain.

Figure 18: In "An abridgement of Kitāb al-masālik wa-al-mamālik by Abū Isḥāq Ibrāhīm b. Muḥammad al-Iṣṭaḵrī" (4/10th century) Or. 3101, Leiden University Library, Digital Collections, Leiden, The Netherlands. Image in the public domain. Use of this resource is governed by the terms and conditions of the Creative Commons CC by license.

Figure 19: From *The Book of Curiosities*, one copy is held at the Bodleian Library MS. Arab. c. 90, Oxford University, Oxford, U.K. Image in the public domain.

Figure 20: From *Nuzhat al-mushtāq fī ikhtirāq al-āfāq*, also known as *The Book of Roger*, Bodleian Library, Oxford University, Oxford, U.K. MS. Pococket 375, folio 3a and 3b. Image in the public domain.

Figure 21: From the manuscript *Kharīdata al-Ajā'ib wa farīdat al gha'rāib* (*The Pearl of Wonders and the Uniqueness of Strange Things*) held at the Library of Congress, Washington, D.C., control number 2013415532, World Digital Library. Image in the public domain.

Figure 22: Piri Reis map of Venice. Image in the public domain.

Figure 23: The *Da Ming Hunyi Tu* is housed at the The First Historical Archives of China, Beijing, China. Image in the public domain.

Figure 24: This image of the Kangnido map was taken by members of Hong Kong Baptist University (Cartography, GEOG1150, 2013, Qiming Zhou *et al.*, Department of Geography, Hong Kong Baptist University). Image in the public domain.

Figure 25: First T-O map by Isidore of Seville, housed at the British Library, London, U.K. Image in the public domain.

Figure 26: The Cotton Map, housed at the British Library, London, U.K.; Cotton MSS, Tib. B. V, folio 56. Image in the public domain.

Figure 27: The Sawley Map, housed at The Parker Library, Corpus Christy College, University of Cambridge, Cambridge, U.K. Image in the public domain.

Figure 28: The Ebstorf Map, colorized photo reproduction. Image in the public domain.

Figure 29: Hereford Map, held at Hereford Cathedral, Hereford, U.K. Image in the public domain.

Figure 30: The Vesconte Map held at British Library, London, U.K. Image in the public domain.

Figure 31: Andrea Bianco's World Map, housed at Biblioteca Nazionale Marciana, Venice, Italy. Image in the public domain.

Figure 32: Reproduced photograph of the now destroyed Walsperger Map, Vatican Apostolic Palatina, Heidelberg, Germany. Image in the public domain.

Figure 33: Genoese Map, housed at Biblioteca Nazionale Centrale di Firenze, Florence, Italy. Image in the public domain.

Figure 34: Leardo World Map, National Geographic Society, house at The University of Wisconsin Libraries, Milwaukee, Wisconsin, used by permission. Image in the public domain.

Figure 35: De La Cosa Map, Naval Museum of Madrid, Madrid, Spain. Image in the public domain.

Figure 36: Waldseemüller map, housed at the Library of Congress, Geography and Map Division, Washington, D.C. Image in the public domain.

Figure 37: Mercator Map, housed at the Maritiem Museum Rotterdam, The Netherlands or University of Basel, Switzerland. Composite image by Wilhelm Kruecken, 201. Image in the public domain.

Figures 38–39: Landsat imagery courtesy of NASA Goddard Space Flight Center and U.S. Geological Survey.

Figure 40: Courtesy of NASA Goddard Space Flight Center.

Notes

1: A SENSE OF PLACE: THE HUMAN URGE TO DRAW GEOGRAPHY

1   (Deacon, 1998)
2   (C. D. Smith, 1987)
3   (Harley, 1987; Leach, 1976)
4   (Lewis, 1987)
5   (Piaget, 1936)
6   (Izard et al., 2009)
7   (DeLoache, 2004)
8   (Premack and Premack, 1983)
9   (Boysen, 1993)
10 (Savage-Rumbaugh, Shanker, & Taylor, 1998)
11 (Matsuzawa, 2001, 2009)
12 (Addessi et al., 2008; Addessi & Visalberghi, 2007)
13 (Call and Tomasello, 2006)
14 (Janmaat et al., 2016)
15 (Piaget and Inhelder, 1948)
16 (Blaut et al., 2003)
17 (Tolman, 1948; Kitchen, 1994)
18 (Brooks et al., 2018)
19 At another site, Jebel Irhoud in Morocco, a cranium previously thought to be Neanderthal fossil is now dated as a three-hundred-thousand-year-old *Homo sapiens*, which suggests that the roots of fully modern humans were widespread in Africa before they moved out to other continents (Hublin et al., 2017).
20 (C. Henshilwood et al., 2004; d'Errico et al., 2005)
21 (Bouzouggara et al., 2007)
22 (Tylén et al., 2020)
23 (C. Henshilwood et al., 2003; Sample, 2018; St. Fleur, 2018)
24 (Jabr, 2014)
25 (Tattersall, 2008)
26 (Rodriguex-Hodalgo et al., 2018)
27 (Hoffmann, Angelucci, et al., 2018; Wong, 2010; Zilhão, 2010)
28 (Hoffmann et al., 2018))
29 (Wraggs Sykes, 2020)
30 (Aubert et al., 2018)
31 There is some controversy about who drew what on Spanish caves. In 2018 a group of paleontologists tested the material of wall drawings in a cave in Spain

and determined the art was older than sixty-five thousand years, which would mean before fully modern humans arrived. If those dates are correct, this discovery would mean Neanderthals were the artists, but this work has yet to be confirmed or suggested for other Neanderthal sites. See (Hoffmann, Standish, et al. 2018).

32 (Smith, 1987)

33 (Smith, 1982)

34 Some have suggested this scenario is not a map but a bunch of motifs with a spotted leopard on top, but it looks like houses and a mountain no matter which way you turn it.

35 (Blumer, 1964; Turconi, 1985)

36 (Turconi, 1985; Turconi, 2001)

37 (Smith, 1982)

38 (Blaut et al., 2003)

39 (Smith, 1987)

40 (Blackmore, 1981; De Hutorowicz, 1911)

41 (Smith, 1987)

42 (Wood, 1993)

43 (Spennemann, 2005; Wise, 1976)

44 (Davenport, 1960)

45 (Lyons, 1928)

46 (Davenport, 1960)

47 (Willford, 1981)

48 (Inuit cartography, 2016)

49 (Rundstrum, 1990)

50 (Rundstrum, 1990)

51 Ivory tusk of a walrus, which was carved by an Eskimo and presented to President William Howard Taft in 1890.

52 (Lyons, 1928)

53 (Norris and Harney, 2014)

2: MAPPING THE WORLD BEFORE FRA MAURO

1 Maps of the world, especially those of the Middle Ages and the Renaissance, have been collectively called mappa mundo, mappae mundi, mappamundo, mappa mundi, and mappamundi. I am choosing to go with the spelling used by historian of early maps Evelyn Edson and call them, in singular and plural, mappamundi. See (Edson, 2007).

2 (P. D. Harvey, 1987)

3 In this chapter I concentrate on world maps and their history and set aside all the other types of ancient maps, such as the Peutinger map from the fourth century CE (during the Roman Empire) because it is essentially a long road map, or the *Madaba Mosaic Map* from 560 CE because it features only the Holy Land. I also ignore ancient celestial maps, such as the *Chinese Dunhuang Star Chart* from 650 CE, and have chosen not to write about globes, since their history is a book all by

itself. So, throughout this book I focus on world maps, but there are any number of good resources about all kinds of ancient maps, such as (Brotton, 2014) or, of course, (Harley & Woodward, 1987). I also exclude floor mosaics, which are few, made for decoration and embellishment, full of allegory rather than land and oceans, and they are usually fragmentary and unsigned (Rappaport & Savage-Smith, 2014).

4   They can also tell of human fakery, or perhaps ambition. For example, the *Vinland Map* was brought forth in 1965 as a pre-Colombian world map, but it turned out to be a twentieth-century fake (Yuhas, 2021).

5   (Edson, 2008)

6   According to map historian Jerry Brotton, it took cuneiform scholars at the British Museum to translate the writing and proclaim this artifact as the first world map (Brotton, 2014).

7   *Imago Mundi* is also the name of the contemporary academic journal that specializes in articles about ancient maps.

8   *The Babylonian Map of the World* is now housed at the British Museum in London. And see (Millard, 1987).

9   (van Duzer, 2013)

10 Clay tablets were often used in the Middle and Near East at this time for communicating all sorts of things; museums and libraries all over the world have collections of these tablets. They were early Post-its, made from common inexpensive material and incised quickly to record a transaction, and most of them were squashed back into mud and formed again as another note. Fortunately, some of these tablets had been fired which hardened them enough to last through the ages. Because they are so small and disposable, clay tablets were also passed around, almost like letters through the post. That also meant they were durable enough to survive many centuries and end up in our contemporary hands.

11 (Aujac, 1987b)

12 (Aujac, 1987b)

13 (Aujac, 1987c)

14 Eratosthenes also calculated the Earth's axial tilt.

15 Later Greek historian Strabo somehow saved some fragments of the map (Aujac, 1987a).

16 (McPhail, 2011)

17 (Aujac, 1987c; Thrower, 2008)

18 (Brotton, 2014)

19 (Aujac, 1987c)

20 (Aujac, 1987a)

21 (Aujac, 1987a)

22 The son-in-law of the first Roman emperor, Augustus Caesar, Marcus Vipsanius Agrippa was a master builder of Roman cities in the late first century BCE. Agrippa was also a geographer and supposedly made a world map for Caesar. This map was completed by others and is long gone, but it seems to be the only Roman world map. Yet some suggest that it was the first map to divide clearly

and purposefully the world into the three known well-delineated continents (Thrower, 2008). The earliest known Roman geographer is Pomponius Mela, who wrote the slim volume "The Situation of the World" (*De situ orbis libri III*) around 43 CE in Latin (today it can be purchased online). Mela divided the world into five zones, north to south, and claimed that three of them were uninhabitable because of temperature. He used the Greek idea of the periplus as a logbook to describe coastlines. Although there are no maps by Mela, others later used his ideas to draw world maps. For example, Pirrus de Noah followed Mela when he drew a rectangular world map in 1414 with land reaching to the edge of the map without the usual enclosing oceans.

23   Both are considered Greeks living and working under the influence of the Roman Empire (Dilke, 1987).

24   (Dilke, 1987) p. 183.

25   (Dilke, 1987)

26   (Brotton, 2014)

27   There was also a third projections but that one never gained any traction (Dilke, 1987).

28   (Dilke, 1987)

29   (Karamustafa, 1992)

30   These 17 seventeen books track the Arabic contribution to science (Rappaport & Savage-Smith, 2004).

31   (Karamustafa, 1992)

32   (Pinto, 2016)

33   (Rappaport & Savage-Smith, 2014)

34   (Tibbets, 1992b)

35   (Rappaport & Savage-Smith, 2014)

36   Al-Khwārizmī was also one of the many scholars hired by Caliph Al-Ma'mūn to construct a world map. This map from the first century was lost for generations but then discovered tucked into an encyclopedia in the Topkapi Sarai Museum in Istanbul, by the historian Sezgin. The Al-Ma'mūn map is based on the knowledge of Arab sailors who had already sailed to China, which makes this map more expansive than maps based on Western mariners because the Arabs had gone further east by 800 CE, and they had more experience exploring oceans and traveling up the east coast of Asia (Tibbets, 1992b).

37   (Tibbets, 1992a)

38   Abū Eshāq Estakrī, better known as al-Istakhri, was another great tenth-century Islamic geographer. He followed the tenants of Balkī when composing his book *Routes of the Realms* (*Ketāb al-masālek wa'l-mamālek*), which included twenty regional maps in color as well as a world map. Al-Istakhri's world map, drawn in about 973, is more land than water. Oriented to the south, the Mediterranean drips off toward the Atlantic Ocean, and other bodies of water, presumably seas, are round blue dots with thin rectangles of blue extended out. Following al-Balkhi's style, the whole map seems more like an artistic representation than a drawing of real land, but the colors are glorious—predominately blue and red on a

white background (Tibbets, 1992a). In 977, the Muslim writer and geographer Ibn Hawqal wrote *Faces of the World* (*Ṣūrat al-'Arḍ*) based on earlier geographic manuscripts, the work of al-Istakhri, and what he had seen or heard on his travels. That book includes twenty-two regional maps and a world map centered on Islamic lands, but it also includes areas such as the steppes of what is now Russia, Tibet, China, Kyiv, West Africa, Egypt, Nubia, and Byzantium. Ibn Hawqal's world map is oriented to the south and is an oval set across two pages and it is full of lines without color and so the land is blank and the water dark. This map might have been drawn following the dictates of Ptolemy's second projection. It's hard to evaluate it since the one held in the Bibliothèque nationale de France in Paris seems to have been redrawn over and over. Also, the Middle Ages were in full swing when this map was made, and so Ibn Hawqal follows other writers in fear-based exaggerated accounts of unknown people and lands. For example, he calls the citizens of Palermo, Sicily, "barbarians."

39   (Pinto, 2016)

40   (Johns & Savage-Smith, 2003)

41   (Rappaport & Savage-Smith, 2004, 2014)

42   (Johns & Savage-Smith, 2003)

43   It was copied by al-Idrīsī and his team a hundred years later or, as some think, the copies of this thirteenth-century maps are copies of al-Idrīsī's copy and then put into this manuscript much later (Johns & Savage-Smith, 2003).

44   (Rappaport & Savage-Smith, 2014)

45   (Johns & Savage-Smith, 2003; Rappaport & Savage-Smith, 2004)

46   (Johns & Savage-Smith, 2003)

47   (Rappaport & Savage-Smith, 2004)

48   It was not just geographers who made maps in the Islamic world. Mahmud al-Kashgari was a linguist who worked on the Turkic language group that includes Turkish, Azerbaijani, Turkmen, Uzbek, Kirghiz, and Yakut. His book about Turkic languages, *Divanü Lügat-it-Türk*, also contains a simple circular world map dated to 1072. This map is oriented to the east, unusual for Islamic maps, but a convention that would be followed for centuries in the East and the West. His map naturally focuses on places in Central Asia that speak Turkic languages, but it broadens out from there. The landmasses are approximate, straightforward, and as usual for Islamic maps, absent of the natural geographic complexity of land formation. This map does include countries such as Japan, China, and Ethiopia. Overall, the features are drawn as graphic representations rather than attempting to record places to scale or anything else. It's a colorful map with blue rivers, yellow deserts, red mountains, and yellow dots for cities, countries, and peoples. Most charming to our modern sensibilities is the place of "the footprint of Adam," located on present-day Sri Lanka, where legend says Adam and Eve went when expelled from the Garden of Eden. Gog and Magog are here too, combining medieval mythology with a bit of religious belief about the Garden of Eden that Muslims and Christians share. See (Maqbul Ahmad, 1992).

49   (Brotton, 2014; Maqbul Ahmad, 1992)

50   (Tibbets, 1992b, 1992a). This school of geography was different than Ptolemy's
     view, and it was founded by geographer Abu Abdullah al-Bakrī in 1068 in his
     Book of Roads and Kingdoms (*Kitāb al-Masālik wa'l-Mamālik*) when he was living
     in what is today Spain.
51   (Maqbul Ahmad, 1992)
52   (Maqbul Ahmad, 1992)
53   (Brotton, 2014)
54   (Tibbets, 1992a)
55   (Tibbets, 1992a)
56   (Ledyard, 1994)
57   (Brotton, 2014)
58   For example, see (Whitmore, 1994) and other chapters in Vol 2, Book 2 of *The
     History of Cartography*.
59   (Scafi, 1999)
60   (Edson, 2007)
61   (Edson, 2007; Woodward, 1987)
62   (Pérez, 2014)
63   (Pérez, 2014)
64   (P. D. Harvey, 1997)
65   (Brotton, 2014)
66   (Westrem, 2001; P. D. Harvey, 1996)
67   Edson also suggests that the idea of putting Jerusalem at the center of a map-
     pamundi might have started with Isidore of Spain when he wrote in his encyclo-
     pedia of 600 CE that Jerusalem was the center of Judea. But Isidore did not mark
     that as such on his T-O world map (Edson, 2007).
68   (Edson, 2007)
69   (Edson, 2007)
70   (Woodward, 1987)
71   (van Duzer, 2014)
72   (Edson, 2007) p. 11.
73   The Roman world map by Agrippa was lost long before the *Hereford Mappa
     Mundi* was made 1,200 years later, but scholars believe that map was the basis for
     the Hereford map, which would underscore a lasting influence of Roman power,
     and cartographic ideas, on later maps of the world (van Duzer, 2014).
74   (Edson, 2007)
75   (Edson, 2007)
76   Forty years later, in about 1342, Ranulf Higden, a German monk at a Benedic-
     tine abbey wrote a long, seven-volume historical manuscript titled *Polychronicon*.
     This book was much admired, and its fame lasted two hundred years and was
     considered accurate history. Within that manuscript is a world map, set on two
     pages. It's oval and oriented to the east with Jerusalem in the center. The ocean
     that circles the world is deep green, the land is off-white, and the accents are
     bright red. Coastal edges are smoothed out and appear as undulating fingers.
     Around the edge is a series of heads and inscriptions (Woodward, 1987).

77   Some have suggested that women—that is, nuns—drew this map (Ven den Hoonaard, 2013).

78   (Edson, 2007)

79   (Edson, 2007)

80   Some historians claim that the large peninsula jutting out from East India on this map is actually South America, and that sort of peninsula in on the Walsperger map as well (Gallez, 1981).

81   This map is now in the Vatican Library.

82   There is also a mysterious German world map that was engraved on a twenty-four-inch-diameter copper plate. It's called the Borgia map after the Italian cardinal who bought that plate from an antique shop in Portugal in 1774. This map is now held in the Vatican Library. Dated about 1450, the Borgia map is oriented south, as is Fra Mauro's. But the provenance of this map is unclear—it must be German because the script is in German; it features comments about the Crusades, Charlemagne in Spain, invading Mongols, and the usual scary fictional savages Gog and Magog at the perimeter of the world. Oddly, the Garden of Eden on this map is at the mouth of the Ganges in India, which is portrayed as a landmass full of marvels and gems. This map sports various types of ships, which also presages Fra Mauro's map. Piero Falchetta, an expert of Fra Mauro's map, believes that some earlier copy of the Borgia map might have been a prototype for Fra Mauro as he set about making his map (Falchetta, 1995, 2006).

83   (Edson, 2007; Sheehan, 2013)

84   (Campbell, 1987)

85   Sheehan suggests otherwise. He feels these portolan charts were also about announcing one's class and education, that they were "worldly goods" used by merchant to show off (Sheehan, 2013).

86   (Thrower, 1972)

87   (Edson, 2007)

88   (Edson, 2007)

89   (Edson, 2007; Thrower, 1972)

90   (Edson, 2007)

91   This atlas is now in the Bibliothèque nationale de France (Thrower, 1972).

92   Interestingly, there is an anonymous portolan chart in the library of the Biblioteca Nazionale Marciana, the same place where Fra Mauro's map hangs, which seems to precede the *Catalan Atlas* and might have been used by the cartographers of the *Catalan Atlas*. Although the country and city of original of the chart is unknown, this chart speaks to the possible connection between Catalan cartographers and Venice (Falchetta, 1994).

93   (Small, 2020) One of his portolan charts of the eastern Mediterranean is from 1311 and so it is one the oldest signed and dated surviving medieval portolan charts.

94   (Edson, 2007)

95   (Edson, 2007)

96   (Edson, 2007)

97   (Ferrar, 2021)

98   (Edson, 2007)

99   There is one in the Biblioteca Capitolare di Verona (Verona Chapter Library)
      with Leardo's signature on the bottom right and the date 1442. This one is stored
      in the lid of a wooden display box, under a plate of glass and accompanied by a
      letter about Prester John, the mythical Christian ruler who sometimes showed
      up on medieval mappamundi along with his made-up kingdom. The copy held
      at Musei Civici Vicenza (Civic Museums of Vicenza) bears the date 1448 and
      although the map format is the same, there are several more rings around the
      world's circumference of this one than the Verona map, and it has two empty red
      circles in the top right and left corners. A third copy was privately donated to the
      American Geographical Society and that one is now at the University of Wis-
      consin Libraries, signed and dated 1450 something since the last digit is smeared.
      Scholars think it reads 1452 or 1453. There might be a fourth Leardo map from
      1447, but that one is lost, if it ever existed (Wright, 1928).

100   The oldest and simplest is in Verona; it has fewer rings around the earth. The
      youngest in Milwaukee is more elaborate—it has ten rings around the earth. This
      one measures 23.75 inches wide (60 cm) and 28.75 inches tall (73 cm) or about
      2 feet (61 cm) in diameter. The map itself is only 15.75 inches (40 cm) in diam-
      eter. The first ring of writing on the Milwaukee Leardo map can be used to
      calculate the date of Easter for ninety-five years; the next ring has the months of
      the year and the signs of the zodiac. The third ring shows the phases of the moon,
      and the fourth through sixths rings carry the days of the months, the hours of
      the day, and points within the hours made up of 1,080 increments, which do
      not exactly correspond to today's seconds. The seventh ring on this expanded
      map presents the Dominical Letter, a system for charting days. The eighth has
      the lengths of the days over the years with the variation of length over the years
      marked on the ninth ring. The last ring lists the saints connected with each day.

101   (Small, 2020)

102   (Woodward, 1987)

103   (Harley, 1987)

104   (Woodward, 1987)

105   (Small, 2020)

106   (Woodward, 1987)

107   (Woodward, 1987)

3: THE WORLD OF FRA MAURO

1   Usually Italian cities have *quartiere*, meaning "quarters" or neighborhoods, and
      most often four of them based on the Latin (*quattuor*) and Italian (*quattro*) words
      for *four*. But Venice has six established neighborhood and so they are called *sest-
      iere*, meaning six (*sei*) neighborhoods.

2   (*Venice's "Bricole": Sentinels of the Lagoon*, 2020)

3   The official name is San Michele in Isola, which refers to the church and monas-
      tery built here in 1212 and named after the Archangel Michael. The *in* translates
      into English as *on* an island.

4   There is also a rumor that the city sold those bones to various industries such a sugar production (Morris, 1960).

5   (*Death and Burial in Venice: What Does the Floating City Do with Its Dead?*, n.d.)

6   In November 2019, the city erected this pontoon bridge for the first time since 1950. For Venetians, it was an historical moment, a memory reenacted. It was also a special moment for those of us who were lucky enough to be there and participate in the walk across the water to honor Venice's dead.

7   (Povoledo, 2018)

8   In 2018 five of the chapels in the older section of the cloister were up for sale because the families that used to be buried there, and who kept up the maintenance of those chapels, were long gone (Povoledo, 2018).

9   (Cazzagon, 2021)

10   (Vigilucci, 1988; Turley, 2006)

11   (Vigilucci, 1988)

12   (Vigilucci, 1988)

13   (Cattaneo, 2011)

14   The name Murano is often added to San Michele as in San Michele in Murano and that has suggested to some that Fra Mauro lived on the island of Murano which is a stone's throw to the north of San Michele. But the monastery of San Michele is not, and never was, on Murano. It is on the island next door (Vigilucci, 1988).

15   (Howard, 2002; Savoy, 2002)

16   Also, turning marble into fake bricks was innovative for the time. Such rustication had been used previously in Florence to give any building a defensive look, like a castle or fortress. Some have suggested Codussi might have used it here as a sign that the island was a hermitage and outsiders were not welcome. The small church is also often touted as a sign of the transition from Gothic to Renaissance style which began later in Venice when compared to the rest of Italy (Howard, *Architectural History of Venice*, 2002) p. 135.

17   Personal communication with Erika Matteo, archaeologist working in Venice, June 2019.

18   (*History of the Venetian Civic Cemetery*, 2016)

19   The vaparetto, or waterbus, stop for San Michele is rightly labeled cimitero, meaning "cemetery" in Italian.

20   (McGreggor, 2006)

21   (McGreggor, 2006)

22   (*History of the Venetian Civic Cemetery*, 2016)

23   (Cross & Livingston, 2005; Vigilucci, 1988)

24   (Bowd, 2002)

25   (Turley, 2006)

26   (Savoy, 2002). This also includes the white scapular, which is a garment that goes over the shoulder and drapes down the front and back and looks somewhat like a hood. It was originally designed as an apron for working monks to keep their garments clean (Cross & Livingston, 2005).

27   (Turley, 2006)

28   Camaldolese monasteries that follow the traditions of the order exist today—
      there are at least thirteen active monasteries around the world. For example, there
      is the original one in Camaldoli, Italy, and another in Big Sur, California. You
      can visit both places and even rent a room for a few days of quiet contemplation.
      See (*New Camadoli Hermitage*, n.d.) or (*Camaldoli, A Hermatige and a Monas-
      tery*, n.d.). There are also eleven active Camaldolese convents and they, too, offer
      retreats (Vigilucci, 1988) and see (*Camaldolese Nuns of Poppi*, n.d.).

29   (Belisle, 2002)

30   (Belisle, 2002)

31   (Hale, 2002)

32   (Belisle, 2002; Hale, 2002)

33   This type of interconnected inter-personal interaction might be viewed as coun-
      tercultural compared to the philosophy of individualism, which is rife in so many
      Western cultures today, especially the United States (Hale, 2002).

34   (Zurla, 1806a) Zurla was also a Camaldolese monk.

35   Mauro is the Italian version of Maurice.

36   (Cattaneo, 2011)

37   (Cattaneo, 2011)

38   (Falchetta, 2013a)

39   (Cattaneo, 2011) p. 19.

40   (Cattaneo, 2011)

41   The engraving plate and the original print are now lost (Cattaneo, 2011).

42   (Cattaneo, 2011)

43   (Cattaneo, 2011)

44   (Falchetta, 2013a; Zurla, 1806a)

45   (Lane, 1933a; Small, 2020; Zorzi, 1999)

46   (Small, 2020)

47   (Kurlansky, 2002; Small, 2020)

48   (Small, 2020)

49   (Falchetta, 2013a)

50   (Small, 2020)

51   (Edson, 2007)

52   (Falchetta, 2013a)

53   This is not punting in the British way where an oar is used as a lever to push the
      boat forward.

54   (Zorzi, 1999)

55   (Zorzi, 1999)

56   (Small, 2020)

57   (Small, 2020)

58   (Cotton, n.d.)

59   (Campagnol, 2014) By the sixteenth century there were fifty convents spread over
      Venice and various islands in the lagoon.

60   (Laven, 2003)

61   (*Monestary Stays*, n.d.)
62   (Bowd, 2002)
63   (Lane, 1973)

4: FRA MAURO MAKES A MAP

1   All the records of the Camaldolese monastery on San Michele have been lost. What we know comes from Placido Zurla's book (available in Italian only) from 1806 (Zurla, 1806b). The record books were lost after that.

2   Double entry bookkeeping was invented in Venice (Small, 2020), and the original method used three different registers, which these lines illustrate.

3   (Cattaneo, 2011)

4   (Cattaneo, 2011)

5   Today these plants are used for ornamentation in gardens or accent in flower arrangements.

6   (Kurlansky, 2016)

7   Note that the contemporary specialized paper used for cooking is called parchment paper, suggesting a combination of parchment and paper, but it obviously is not a combination of animal skin and paper.

8   Interestingly, the same process is used for blocking knitted fabric. Once an item is knitted with yarn—that is out of wool that is sheep's hair—it is also washed to remove any leftover bits of dirt from the animal or the manufacturing process. The garment is then stretched wet onto a frame to the size and shape desired. As it dries, the yarn blossoms, fluffs out, and the garment retains the shape of the frame, even after removal. The difference between knitted woolen garments and parchment is that you can rewash and block a sweater, but you can't do that with a painted parchment. Both are making use of domestic animal, but in the case of yarn, no animals are killed, they are only shorn.

9   (Ryder, 1964)

10   The invention of paper is probably one of the most universally significant, and longest-lasting, technologies ever invented (Kurlansky, 2016). From China, it was introduced into the Islamic world in the eighth century and onto Europe by the eleventh century. Damage to paper maps could be caused by direct sunlight; candlelight, with its spiral of smoke and ash that drifts through the air; and the smoke and ash of room fires used as indoor heat during the Middle Ages.

11   This video was made when the map was loaned to the National Library of Australia in 2013 for an exhibit. It was the first time the map ever left Venice. The video gives an inside look into how it was made and how it fits together, and how big it is (National Library of Australia, 2014).

12   (*Tools of the Trade Part 1*, 2010)

13   Although fresco was the most common way to paint during the late Middle Ages, Venice quickly opted for oil paint on canvas because fresco simply fell off the walls of Venetian buildings. The waters of the lagoon notoriously seeped up walls leaving damaging salt deposits in their wake which undermined fresco. And so most Venetian artists were advocates of oil paint, a medium they learned

from Flemish painters. But oil paint was expensive, and it took a very long time to dry, and so tempera and ink were good options.

14 (Small, 2020)
15 (Special Collection Conservation Unit, 2012)
16 (*How to Make Ink in the Middle Ages*, n.d.)
17 (Hildago et al., 2018)
18 (*Medieval Age (500–1400)*, n.d.)
19 (*Tools of the Trade Part 1*, 2010)
20 (Stewart, 2018)
21 (The Colorful Medieval World: Mixing Ancient Colors, 2019)
22 (*Tools of the Trade Part 1*, 2010)
23 (Cattaneo, 2011)
24 (Edson, 2007)
25 (Cattaneo, 2011)
26 (Cattaneo, 2011)
27 (Small, 2020)
28 (Cattaneo, 2011)
29 (Cattaneo, 2011)
30 (Edson, 2007) p. 151.
31 (Vogel, 2011)
32 (Small, 2020)
33 Venetian Giovanni Battista Ramusio complied three books of travel stories in the mid-1500s, and he claimed that Marco Polo brought back a world map from his travels, and that Fra Mauro had access to that map. There is no confirmation of this claim (Falchetta, 2006; Ramusio, 1606).
34 (Small, 2020)
35 (di Robilant, 1011)
36 (Falchetta, 2006)
37 (Falchetta, 2006)
38 (Edson, 2007)
39 (Falchetta, 2006)
40 (Cattaneo, 2011)
41 (Cattaneo, 2011)
42 (Brentjes, 2015)
43 (Alamgià, 1944; Ratti, 1988; Winter, 1962)
44 (Alamgià, 1944)
45 (Cattaneo, 2011)
46 (Falchetta, 2006) inscription 2489, and (Mauntel, 2018)
47 (Mauntel, 2018)
48 (Mauntel, 2018)
49 (Mauntel, 2018)
50 (Mauntel, 2018)
51 (Edson, 2007)
52 (Falchetta, 2006) inscription 1312.

53   (O'Doherty, 2011)

54   (Cattaneo, 2011) Also note that the word *spice* was used in Venice to refer to all
      manner of luxury items that had to be imported from far away.

55   (Falchetta, 2006)

56   (Cattaneo, 2015); see also (O'Doherty, 2011) for a discussion on the Indian
      Ocean in particular.

57   (Cattaneo, 2011) p. 112.

58   (Edson, 2007)

59   (Edson, 2007; Falchetta, 2006)

60   (Edson, 2007)

61   (Cattaneo, 2011)

62   (Cattaneo, 2011)

63   (Edson, 2007)

64   (Edson, 2007)

65   (Woodward, 1985)

66   (O'Doherty, 2011)

67   (Edson, 2007; O'Doherty, 2011)

68   (Edson, 2007)

69   (Edson, 2007) p. 150.

70   (Cattaneo, 2011)

71   (Falchetta, 2006) p. 122.

72   (Edson, 2007)

73   (Cattaneo, 2011)

74   (Cattaneo, 2011)

75   (Cattaneo, 2011)

76   (Edson, 2007)

77   (Cattaneo, 2003; Macron, 2001, 2006)

78   (Falchetta, 2006)

79   (Cattaneo, 2003) p. 99.

80   Legend on Fra Mauro's map, lower left corner. See (Falchetta, 2006) and (Cat-
      taneo, 2011).

81   (Edson, 2007) p. 146.

82   (Cattaneo, 2011; Falchetta, 1995)

83   (Cattaneo, 2011)

84   (Falchetta, 2013b)

85   (National Library of Australia, 2013, 2014)

86   In another room is a large screen for an interactive website about the map recently
      developed by the Museo Galileo in Florence (*Fra Mauro's World*, n.d.).

5: WHAT FRA MAURO WANTED TO TELL US ABOUT GEOGRAPHY

1   For this chapter I rely almost exclusively on two essential and unique books: Piero
     Falchetta's *Fra Mauro's World Map*, which includes the translation of the three
     thousand inscriptions into English, and Angelo Cattaneo's *Fra Mauro's Mappa
     Mundi and Fifteenth Century Venice*. Together they comprise the most thorough

description, analysis, historical context, and historical significance of this map (Cattaneo, 2011; Falchetta, 2006). Therefore, they are cited repeatedly here.

2   (Falchetta, 2006)

3   (Cattaneo, 2011)

4   (Cattaneo, 2011)

5   (Cattaneo, 2011) pgs. 268–69.

6   (Cattaneo, 2011) p. 270.

7   (Falchetta, 2006) inscription 1888.

8   (Cattaneo, 2011)

9   That tradition continues with the learned and informed analyses of this map by Andrea Cattaneo and Piero Falchetta as they worked through both the cartography and the inscriptions (Cattaneo, 2011; Falchetta, 2006).

10 (Scafi, 1999)

11 (Cattaneo, 2011) p. 89.

12 (Falchetta, 2006) inscription 2193.

13 (Cattaneo, 2011)

14 (Falchetta, 2006) inscription 1000.

15 (Cattaneo, 2011)

16 (Falchetta, 2006) inscription 389.

17 (Falchetta, 2006) inscription 2202.

18 The Italian we speak today is a version of the Tuscan language of the great fourteenth-century Italian writers: Dante, Petrarch, and Boccaccio. But these authors, especially Dante, wanted their work to be read by the masses, and so they wrote in their home languages, which were called *vulgate*, meaning "for the common man." The only reason common Italian today is the Tuscan variant is that state officials decided to adopt it the national language after Italy was unified as a kingdom, a process that was not fully complete until 1918. While Tuscan Italian is now taught in schools and used in business, people of the various regions continue to speak in their native languages with each other. These dialects, or separate languages, are alive and kicking in the streets and towns of the various regions.

19 (Pizzati, 2007)

20 (Small, 2020)

21 (Pizzati, 2007)

22 In the last few years, the Venetian government, often made up of people who are not Venetians, have repeatedly tried to replace Veneziano with Italian to make life easier for tourists, many of whom come from other areas of Italy. One project set out to change the words or the spelling of words on street signs, which are painted high up on the sides of buildings in Venice. The standard sign is a white rectangle edged in black with the street name in black letters. The government went about wiping out Veneziano names and inserting double letters in some words to make them look more like Italian. In the end, the bastardized Venetian-Italian words made no sense to anyone. The backlash was enormous from locals and soon many signs sported both Venetian and Italian names as a

concession. And in the style of graffiti protest, someone who became known as the Bandit of San Polo (one of the neighborhoods of Venice) went out at night with a ladder and a paintbrush blocking out all the double letters on the street signs and turning them back into Veneziano. Venetians resisted because their language is a point of identity and trying to wipe it out for the benefit of outsiders is unacceptable.

23   A communal identity was also behind the use of the lion of St. Mark all around town on buildings and flags as a constant reminder of the collective. Indeed, the Fra Mauro map was called a "treasure" of Venice in an edited volume of travel writing by Giovanni Ramusio in the sixteenth century and the fact that it was drawn by a Venetian monk, hung in Venice, and was written in Veneziano would underscore that honor (Small, 2020; Wills, 2001; Zorzi, 1999).

24   (Brotton, 2014)

25   (*Fra Mauro's World*, n.d.)

26   (Falchetta, 2006) inscription 2892.

27   (Falchetta, 2006) inscription 1405.

28   (Falchetta, 2006)

29   (Falchetta, 2006) inscription 2834.

30   (Falchetta, 2006) p. 123.

31   (Falchetta, 2006) p. 123.

32   (Falchetta, 2006) p. 133.

33   (Falchetta, 2006)

34   (Falchetta, 2006)

35   Marco Polo's journey was written down by Rustichello da Pisa and Niccolò de Conte's was written down by Florentine humanist Poggio Bracciolini.

36   There was a suggestion that Marco Polo had maps when he came back from China, which certainly makes sense, but his book had none and there is no evidence of these maps.

37   (Falchetta, 2006) inscription 2719.

38   (Falchetta, 2006) inscription 1312.

39   (Falchetta, 2006) inscription 2489.

40   (Falchetta, 2006) inscription 2847.

41   It's impossible to know if this story is true or if it is just a fable or fantasy, but these tales were repeated in Strabo's *Geographica* in the first century, and he was an accomplished explorer as well and Strabo might have been able to check the veracity of these stories.

42   (Lane, 1933b, 1934; Small, 2020)

43   But the cape did appear on some nautical charts in the late fourteenth century (Falchetta, 2006).

44   (Falchetta, 2006) inscription 149.

45   (Falchetta, 2006) inscriptions 19 and 319.

46   (Falchetta, 2006) inscription 149.

47   (Falchetta, 2006) inscription 53.

48   (Falchetta, 2006)

49   (Falchetta, 2006) inscription 562.
50   This idea was also repeated by another Islamic geographer, Ibn Saidin, in the thirteenth century, and others.
51   (Falchetta, 2006) inscription 49.
52   (Falchetta, 2006; Ramusio, 1606)
53   (Zurla, 1806)
54   (Falchetta, 2006)
55   (Falchetta, 2006)
56   (Falchetta, 2006) inscription 98.
57   (Falchetta, 2006)
58   (Falchetta, 2006) inscription 480.
59   (Falchetta, 2006)
60   (Falchetta, 2006) inscriptions 282, 416, and 589, for example.
61   (Falchetta, 2006)
62   (O'Doherty, 2011) p. 32.
63   (Falchetta, 2006) inscriptions 51 and 231.
64   (Falchetta, 2006) inscriptions 594 and 1351.
65   (O'Doherty, 2011)
66   (Falchetta, 2006) inscription 76.
67   (Falchetta, 2006) inscription 416.
68   (Falchetta, 2006) inscription 240.
69   (Falchetta, 2006) inscription 24.
70   (Falchetta, 2006) inscription 25.
71   (Falchetta, 2006) inscriptions 264 and 282.
72   (O'Doherty, 2011)
73   For a comparison of the toponyms Fra Mauro used for South Asia with other maps see (Simon, 2017).
74   (Falchetta, 2006)
75   (Falchetta, 2006) inscription 1422.
76   (Falchetta, 2006) inscription 1742.
77   (Falchetta, 2006) inscription 793.
78   (Falchetta, 2006) inscription 957.
79   (Falchetta, 2006) inscription 960.
80   (Falchetta, 2006) p.90.
81   (Falchetta, 2006) inscription 1179.
82   (Falchetta, 2006) inscription 1188.
83   (Falchetta, 2006) inscription 1863.
84   (Falchetta, 2006) inscription 1199.
85   (Falchetta, 2006) inscription 1264.
86   (Falchetta, 2006) inscription 2697.
87   (Falchetta, 2006) inscription 2702.
88   (Falchetta, 2006) inscription 2699.
89   (Falchetta, 2006) inscriptions 2698 and 2701.
90   (di Robilant, 1011; Small, 2020)

91   There is a statue of Querini and his sled dogs in Venice at one entrance of the
      Giardini, and a city rowing club named after him.
92   (O'Doherty, 2011)

6: PEOPLES, GOODS, MYTHS, AND MARVELS: LESSONS FROM FRA MAURO

1   (Nanetti et al., 2015)
2   (Falchetta, 2006) inscription 48.
3   There are many translations of this book into many languages. I used one printed
      in 2012, *The Travels of Marco Polo*, as it is often titled in English language ver-
      sions. The publisher notes that this is a reprint of a translation by William
      Marsden (1754–1836), which is based on Venetian Giambattista Ramusio's
      printed edition of 1553, which was in Italian. The publisher also notes that the
      Marssen translation was the first major modern English translation of this work
      and remains so. The author is listed as Marco Polo.
4   These were Marco Polo's formative years and he spent them in the service of the
      Great Khan integrating himself into a culture that was entirely different than
      Venice. Polo was such a star that the Khan used him for diplomatic missions and
      made him an emissary to various parts of his vast kingdom. On those trips, Polo
      was a diplomat and observer, not a merchant, as he traveled to Sri Lanka, Indonesia,
      Vietnam, and all around China. He also visited the Middle East and the Levant
      with his father and uncle when they were going and coming from Venice. In total,
      they were away from Venice for twenty-four years. In the introductory chapter of
      Il Milione, Rustichello da Pisa writes vividly about Marco Polo and his acceptance
      at the court of Kublai Kahn, the Mongol emperor who ruled present-day China.
      "Marco was held in high estimation and respected by all belonging to the court. He
      learned in a short time and adopted the manners of Tartars [Mongols], and acquired
      proficiencies in four different languages, which he became qualified to read and
      write" (Polo, 2012). The only difference between Marco Polo and a modern-day
      anthropologist is that he stayed a lot longer than the year or two that most academics
      put into living with the people they want to describe and understand.
5   (Cattaneo, 2015)
6   Taking that method even further, a recent project used Google Maps to align
      some points of interest on the map and in Polo's book to the contemporary
      landscape, and they found that he was most often correct in his descriptions
      and locations (*Fra Mauro's Map of the World (Dated 26 August 1460)*, n.d.) and
      see chapter 8.
7   (Falchetta, 2006) inscription 317.
8   (Falchetta, 2006) inscription 1437.
9   (Falchetta, 2006) inscription 2711.
10   (Falchetta, 2006) inscription 2240.
11   (Falchetta, 2006) inscription 2240.
12   (Falchetta, 2006) inscription 2357.
13   (Falchetta, 2006) inscriptions 2279 and 2282.
14   (Falchetta, 2006) inscription 317.

15   (Falchetta, 2006) note near the island of Crete.

16   (Falchetta, 2006) inscription 1199.

17   (Falchetta, 2006) inscription 28.

18   (Falchetta, 2006) inscription 2697.

19   (Falchetta, 2006) inscription 2315.

20   (Falchetta, 2006) inscriptions 2315 and 2317.

21   (Falchetta, 2006) inscription 1451.

22   (Falchetta, 2006) inscription 933.

23   (Falchetta, 2006) inscription 448.

24   (Falchetta, 2006; *Hampi: City of Ruins*, n.d.)

25   (Falchetta, 2006) inscription 2729.

26   (Falchetta, 2006) inscriptions 2734 and 2508.

27   (Falchetta, 2006) inscription 166.

28   (Falchetta, 2006) inscription 2733.

29   (Falchetta, 2006) inscription 2719.

30   (Falchetta, 2006) inscription 650.

31   (Falchetta, 2006) inscription 771.

32   (Falchetta, 2006) inscription 920.

33   (Falchetta, 2006) inscription 2706.

34   (Falchetta, 2006) inscription 1128.

35   (Falchetta, 2006) inscription 2461.

36   (Falchetta, 2006) inscription 2894. Also, in Russia, he marks the place were
      where Timur, the great Mongol leader, once slaughtered millions to defeat the
      Tartars and become one of the greatest rulers in the Islamic world (Falchetta,
      2006, inscription 2526).

37   (Falchetta, 2006) inscription 2323.

38   (Falchetta, 2006) inscription 560.

39   (Falchetta, 2006) inscription 2360.

40   (Falchetta, 2006) inscription 2388.

41   (Falchetta, 2006) inscription 2464.

42   (Falchetta, 2006) inscriptions 2433 and 2434.

43   (Falchetta, 2006) inscriptions 792, 2315, and 2829.

44   (Falchetta, 2006) inscription 2721.

45   (Falchetta, 2006) inscription 865.

46   (Falchetta, 2006) inscription 19.

47   (Falchetta, 2006) inscription 1342.

48   (Falchetta, 2006) inscription 1381.

49   (Cattaneo, 2004, 2015)

50   (Cattaneo, 2015)

51   (*Fra Mauro's World*, n.d.)

52   (di Robilant, 1011)

53   That same addictive lure of spices can be felt, or smelled, today in the long-
      standing spice shop Mascari near Venice's Rialto Market. The windows of
      Mascari are filled with pyramids of spices from all over the world and walking

inside the shop smells like passing a spice stall in an outdoor market in Bali, Java, or some other island where spices are still grown and exported. The experience moves from smell to taste as, say, 100 one hundred grams of cinnamon powder is wrapped up in Mascari paper and brought home and added to cookies, cakes, or hot chocolate. There is no doubt that Mascari powdered cinnamon is better than anything bought in a glass or plastic container in a grocery store because it is so fresh—maybe not right off a trade boat in Marco Polo's time, but still fresh.

54   (Falchetta, 2006) inscription 178.
55   (Falchetta, 2006) inscription 589.
56   (Falchetta, 2006) inscription 23.
57   (Falchetta, 2006) inscription 407.
58   (Falchetta, 2006) inscription 27.
59   (Small, 2020)
60   (Falchetta, 2006) inscription 280.
61   (Kurlansky, 2002)
62   (Kurlansky, 2002; Small, 2020)
63   (Falchetta, 2006) inscriptions 2370, 2419, 2425, and 2723.
64   (Falchetta, 2006) inscriptions 28, 149, 715, and 1347.
65   (Falchetta, 2006) inscription 594.
66   (Falchetta, 2006) inscription 28.
67   (Falchetta, 2006) inscription 2376.
68   (Falchetta, 2006) inscription 326.
69   (Falchetta, 2006) inscriptions 594 and 602.
70   (Falchetta, 2006) inscription 1199.
71   (Falchetta, 2006) inscription 2273.
72   (Falchetta, 2006) inscription 792.
73   (Falchetta, 2006) inscription 205.
74   (Falchetta, 2006) inscription 280.
75   (Falchetta, 2006) inscription 333.
76   (Falchetta, 2006) inscription 4.
77   (Falchetta, 2006) inscriptions 16, 27, 28, 401, 564, 653, and 1426.
78   (Falchetta, 2006) inscription 1426.
79   (Falchetta, 2006) inscriptions 594 and 2323.
80   (Falchetta, 2006) inscriptions 407, 712, 733, and 2254.
81   (Falchetta, 2006)
82   (Falchetta, 2006) inscription 1344.
83   (Falchetta, 2006) inscription 2913.
84   (Falchetta, 2006) inscription 2920.
85   We have recently understood that Ukraine is the breadbasket of Eastern Europe, China, and Africa and that the invasion of Ukraine by Russia in 2022 has caused widespread famine in those receiving countries because they are now dependent on Ukrainian imports.
86   (Falchetta, 2006) inscription 815.
87   (Falchetta, 2006) inscriptions 27 and 792.

88   (Falchetta, 2006) inscription 917.
89   (Falchetta, 2006) inscription 23.
90   (Falchetta, 2006) inscription 1257.
91   (Falchetta, 2006)
92   (Falchetta, 2006) inscription 90.
93   (Falchetta, 2006) inscription 2348.
94   (Falchetta, 2006) inscription 452.
95   (Falchetta, 2006) inscription 779.
96   (Falchetta, 2006) inscription 815.
97   (Falchetta, 2006) inscriptions 27 and 28.
98   (Falchetta, 2006) inscription 178.
99   (Falchetta, 2006) inscription 27.
100 (Falchetta, 2006) inscription 2708.
101 See, for example, (Lee & Daly, 1999)
102 His inscriptions have the same sort of respect that anthropologists strive for when encountering, living with, and studying foreign peoples. We anthropologists have been trained to embrace the idea of cultural relativity, the notion that all cultures are equal even if they appear very different from the Western view. Fra Mauro has a similar sense of cultural relativity, a certain respect for other cultures, but it was often tempered by fear, that medieval horror of the unknown. In any case, he was always trying to report the facts with a certain respect and curiosity. He also exhibited a deep understanding that a diversity of people can affect how a culture operates, and he seemed to revel in the plethora of cultures that humanity has produced. He even encouraged the viewer to know more.
103 (Falchetta, 2006) inscription 1380.
104 (Falchetta, 2006) inscription 2299.
105 (Falchetta, 2006) inscription 779.
106 (Falchetta, 2006) inscription 1382.
107 (Falchetta, 2006) inscription 2480.
108 (Falchetta, 2006) inscriptions 1808 and 2480.
109 (Falchetta, 2006) inscription 820.
110 (Falchetta, 2006) inscriptions 451 and 1329.
111 This information he received from reading the travels of Niccolò de' Conte (Falchetta, 2006) inscription 23.
112 (Falchetta, 2006) inscription 1607.
113 (Falchetta, 2006) inscription 815.
114 (Falchetta, 2006) inscription 695.
115 (Falchetta, 2006) inscription 63.
116 (Falchetta, 2006) inscription 624.
117 (Falchetta, 2006) inscription 249.
118 (Falchetta, 2006) inscription 2414.
119 (Falchetta, 2006) inscription 881.
120 (Falchetta, 2006) inscription 1129.
121 (Falchetta, 2006) for example inscription 885.

122   (Falchetta, 2006) inscription 6.
123   (Falchetta, 2006) inscription 179.
124   (Falchetta, 2006) inscription 195.
125   (Falchetta, 2006) inscription 220.
126   (Falchetta, 2006) inscription 2897.
127   (Falchetta, 2006) inscription 27.
128   (Falchetta, 2006) inscription 2480.
129   (Falchetta, 2006) inscription 1054.
130   (Falchetta, 2006) inscription 31.
131   (Falchetta, 2006) inscription 2910.
132   (Larner, 1999; Polo, 2012)
133   (Falchetta, 2006) inscription 2697.
134   (Falchetta, 2006) inscription 2708.
135   (Falchetta, 2006) inscription 1467.
136   (Falchetta, 2006) inscription 2649.
137   (Falchetta, 2006) inscription 2880.
138   (Falchetta, 2006) inscriptions 2880 and 2921.
139   (Falchetta, 2006) inscription 152.
140   (Falchetta, 2006) inscription 208.
141   (Falchetta, 2006) inscription 579.
142   (Falchetta, 2006) inscription 27.
143   (van Duzer, 2013)
144   (Falchetta, 2006) inscription 1457.
145   (Falchetta, 2006) inscription 2817.
146   (Falchetta, 2006) inscription 1043.
147   (Falchetta, 2006) inscription 707.
148   (Falchetta, 2006) inscription 2907.
149   (van Duzer, 2013) p.57.
150   (Falchetta, 2006) inscription 208.
151   (Falchetta, 2006) inscriptions 208 and 2212.
152   (Falchetta, 2006) inscription 385.
153   (Falchetta, 2006) inscription 460.
154   (Falchetta, 2006) inscription 28.
155   (Gow, 1998)
156   For example, (Scherb, 2002; van Donzel & Schmidt, 2009)
157   (Falchetta, 2006) inscription 2752.
158   (Falchetta, 2006) inscription 2403.
159   (Falchetta, 2006)
160   (Silverberg, 1972) among many historical analyses.
161   (Falchetta, 2006) inscription 57.
162   (Falchetta, 2006) inscription 66.
163   (Falchetta, 2006) inscription 77.
164   (Falchetta, 2006) inscription 83.
165   (Cattaneo, 2011)

7: THE CONSEQUENCES OF FRA MAURO'S MAP

1   It's also possible that European traders saw an opportunity for customers in all those other cultures and people.

2   (Cattaneo, 2011)

3   (Wilford, 1981)

4   (Wilford, 1981)

5   (Cattaneo, 2011)

6   (*Fra Mauro's World*, n.d.)

7   (Cattaneo, 2011)

8   (Falchetta, 2006)

9   (Cattaneo, 2011) p. 324.

10 The Sala dello Scudo is open to visitors as they walk through the many rooms of the doge's apartments at the Palazzo Ducale.

11 (Cattaneo, 2011) p. 310.

12 (Cattaneo, 2011) p. 324.

13 About the same time, from 1519 to 1524, another Venetian named Alessandro Zorzi, was compiling four notebook of travel diaries of others that can best be called guidebooks. He, too, wrote about Fra Mauro's map and compiled an index of the map. Zorzi apparently had access to pages from Fra Mauro's original notes for the map before they disappeared. Interestingly, Zorzi referred to Mauro as "Frate Nicola da S. Michiel di Muran," which is confusing, especially since those original notes no longer exist and no one knows if or with what name Fra Mauro signed those notes (Cattaneo, 2011). For all we know, Zorzi was using Fra Mauro's real first name, Nicola, or he simply got it wrong. In any case, Zorzi's quoting Fra Mauro's notes is significant because it means the monk was considered a reliable source for geography, at least by a fellow Venetian monk. At the same time, Zorzi is very confused about notes on Ethiopia that appear on the map and attributes them to a completely different monk who he claims had visited Ethiopia in 1470, but that was after the map was long finished. The Ethiopian notes, as Mauro explains on the map, came from Ethiopian clerics who had visited San Michele and spoke with Fra Mauro. All this would suggest that Zorzi was confused about Fra Mauro's role in this map, and the timeline of its production.

14 (Cattaneo, 2011)

15 (Cattaneo, 2011)

16 (Cattaneo, 2011; Falchetta, 2006)

17 (Cattaneo, 2011)

18 You can still buy this slim volume today (Zurla, 1806). Also, Zurla's treatise was helped by scholarly work on the map that was never published by fellow Camaldolese monk, Bartolomeo Alberto Cappellari, who went on to be Pope Gregory XVI. Zurla also wrote a book about the explorations of Marco Polo. He could have risen in the hierarchy of the Catholic Church and was offered many opportunities to do so, but Zurla regularly them turned down. He did, however, eventually become a cardinal.

19   It is cataloged as "additional MS 11267 (Plate IX)"

20   (Cattaneo, 2011)

21   (Cattaneo, 2011) p. 324.

22   (*Mappamundo Drawn by Fra Mauro, 1459 [Cartographic Material]*, n.d.)

23   (*Life-Size 1871 Photograph of the Fra Mauro Map of the World-Carlo Naya*, n.d.).
      Or you can simply buy a facsimile or poster of the image at fineartamerica.com.

24   (Falchetta, 2013b; Leporace, 1956)

25   (Cattaneo, 2011)

26   Although not known well by most of the world, Fra Mauro's map is considered
      precious to those who know the story of maps. It left Venice for the first time
      in 2013, when it removed from the wall, crated, and flown to Australia for an
      exhibit on the history of maps at the National Library in Canberra. How inter-
      esting that this map was taken to a continent that was known only to indigenous
      peoples back when it was drawn, a continent that had yet to appear on world
      maps. Videos of the transfer (National Library of Australia, 2014) and a discus-
      sion of the map (National Library of Australia, 2013) are fun to watch because
      you get a feel for the size of the map and how it was put together.

27   (Morse, 2007)

28   (Scardia et al., 2021)

29   What makes exploring the moon and Mars so interesting is that unlike previous
      human explorations on Earth, others hadn't gone before.

30   (Magno, 2013; Small, 2020)

31   For a discussion of projection and other world maps from the Renaissance not
      mentioned here, see (Snyder, 2007).

32   (Dalché, 2007). Also, at this point in cartographic history, maps were being
      printed, which was, of course, a major technological achievement and opportu-
      nity. In 1482 Ptolemy's *Geography* was printed in Ulm, Germany, and it included
      a woodblock print of Ptolemy's instruction for a world map by Johannes Schnitzer
      and this map also included Greenland.

33   Note that I will focus on the best-known and influential world maps at this
      point and not include lesser known or more fragmentary works. See also notes
      44, 51, and 59.

34   (Wilford, 1981)

35   (Wilford, 1981)

36   (Wilford, 1981)

37   John Noble Wilford, the author of *The Mapmakers*, says that the Portuguese
      thought that Christopher Columbus was a fool to assume he had landed on Asian
      islands, and yet the treaty had given Spain, not Portugal, the right to that land as
      they set out not just to explore but also conquer and exploit the New World.

38   (Brotton, 2014)

39   (Brotton, 2014)

40   The de la Cosa mixture of portolan and mappamundi is also the first to show
      the equator, which the Portuguese had passed in 1473, and the Tropic of Cancer
      as red lines across the world. It also includes a new gray line north to south that

codified the Treaty of Tordesillas of 1494. That line was a sort of peace treaty between Spain (Castile) and Portugal, the two big players in the Age of Exploration. In agreement, it divvied up any future discovery of islands east or west of meridian 370, which runs through the Cape Verde islands, between the two nations. This treaty was imperial, suggesting that only Spain or Portugal, and no one else, even the inhabitants of those islands, could "own" places where they happened to land. It was also quickly meaningless once everyone realized that Columbus had not just discovered a few islands but had, instead, stumbled on new continents.

41   (Wilford, 1981)

42   (Brotton, 2014; Wilford, 1981)

43   In 1756 another polar projection world map was part of the book *Marfetname* (Book of Gnosis) by Turkish scholar, astronomer, and Sufi saint Ibrahim Hakki Ezurim. The book combines astronomy, math, anatomy, philosophy, and mysticism. The world map included as an illustration is a two-hemisphere map, one from the perspective of the North Pole and the other from the South Pole. See (Islamic Cosmology and Astronomy: Ibrahim Hakki's Maarifetname, 2010).

44   (Garfield, 2013)

45   (Kurlansky, 1997)

46   (Brotton, 2014; Wilford, 1981)

47   (Steins, n.d.)

48   All Peri Reis maps are gloriously beautiful. For example, his maps of Venice sparkle, delight, and invite, like all great works of art.

49   (Wilford, 1981) p. 86.

50   An Islamic world map dated 1567 has deep and confusing connections to Venice while that city was still one of the several European centers for map printing. That world map is heart-shaped and covered in Turkish writing. It has been historically attributed to Tunisian Hajji Ahmed, but he turns out not to be a real person. Instead, this map might have been made by Venetians, perhaps even Venetian cartographer Giacomo Gastaldi, or someone else, and intended to court the Ottoman map market (Arbel, 2002; Emiralioglu, 2013). The giveaway is that some of the Turkish writing is misspelled. We know the map was a Venetian production because its original woodblocks were rediscovered in the Palazzo Ducale of Venice, the home of the doge and the seat of government, in 1795. The commercial mapmaking business in Italy, especially in Venice, had declined, and there was also competition for printed maps because of the cross-pollination of cartographers and customers as the world connected and became global in the manufacture and construction of goods. The rediscovery of the woodblocks came right before the fall of the Republic of Venice, when the city was scrambling for income, and the current Doge immediately ordered a printing of twenty-four copies. Two of them now belong to the Marciana Library in Venice and one is currently displayed in the exhibit that leads to Fra Mauro's map. Another is a 1567 world map that is part of an atlas attributed to Ali Macar Reis. But Macar Reis was a Hungarian who was working for the Ottomans, the portolan charts

for the atlas were made in Italy, and the world map is a copy of the Venetian
Gastaldi world map made six years earlier in Venice (Emiralioglu, 2013).

51   (Brotton, 2014) p. 103.
52   (Brotton, 2014)
53   Gerard Mercator was also the first to use the word *atlas* for a collection of maps (Brotton, 2014).
54   (Mercator, 1538; Wilford, 1981)
55   (Wilford, 1981)
56   (Wilford, 1981)
57   (Wilford, 1981) p. 104.
58   Joan Blaeu also authored an eleven-volume atlas, filled with maps, and it was huge, still considered one of the largest sized books ever printed.
59   (Kagan & Schmidt, 2007)
60   (Brotton, 2014)
61   In 1989 artist Alighiero Boetti created *Mappa*, (some of which is at the Museum of Modern Art in New York) is an installation of the world with flags on their countries embroidered on linen by Afghani women. Said Boetti, "The world is shaped as it is, I did not draw it; the flags are what they are, I did not design them. In short, I created absolutely nothing."
62   (Scharping, 2016)
63   (Brotton, 2014; Wilford, 1981) and see (*The Peters Projection Map*, n.d.)
64   "Least Distorted Equal Area World Map 2016."
65   (Sokol, 2021)

8: WHY WORLD MAPS STILL MATTER

1   (Small, 2020)
2   (Garner, 2021)
3   (Wilford, 1981)
4   (Wilford, 1981) p. 420.
5   (*Fra Mauro's Mappamundi*, 2014)
6   (*Fra Mauro's Mappamundi*, 2014)
7   (Mai, 2017)
8   (Brotton, 2014)
9   (*Sea Rise Levels and the Fate of Coastal Cities*, n.d.)
10   (Nanetti et al., 2015)
11   (*Fra Mauro's Map of the World (Dated 26 August 1460)*, n.d.) see also (Nanetti, 2022a, b).
12   (Cheon et al., 2016; Nanetti & Cheon, 2016)
13   (*Engineering Historical Memory*, n.d.)
14   The photographs in the various spectrums were made by professional photographer Francesco Mangiaracina during another renovation in 2009. Mangiaracina placed a grid before the map and divided it into twenty-four sections for better resolution (Nanetti & Benvenutti, 2019).
15   (Nanetti & Benvenutti, 2019)

16   (Nanetti, 2024)
17   (Nanetti & Benvenuti, 2021)
18   (Routley, 2021)
19   (D. Smith, 2021)
20   (Brotton, 2014; Maps, n.d.)
21   (*The Decolonial Atlas*, n.d.) There is also a Facebook page.
22   The name was first proposed by J. H. Mädler who had, along with W. Beer, cre-
     ated the *Mappa Selenographica*, a 37-inch-diameter mappamundi of the moon,
     around 1836.
23   See the movie *Apollo 13*, which shows their crisis, the aborted mission, how the
     astronauts and engineers at NASA in Houston solve it, and their spectacular
     return to Earth.
24   (*Apollo 14 Hike to Cone Crater*, 2021)
25   (Uri, 2021) Some of the seeds were rendered useless when that bag burst during
     the decontamination process; the rest were planted. There was a moon tree in
     a park nearby me in Philadelphia, but when I searched it out, the tree had died
     some years ago, but the plaque set into a boulder remains. I had passed it thou-
     sands of times before but had no idea it was so significant. You can search for
     moon tree in your areas with this site https://nssdc.gsfc.nasa.gov/planetary/lunar
     /moon_tree.html (Williams, 2022).

Bibliography

Abbott, Lyman. "More Roseveltiana." *The Outlook*, April 29, 1925. Fordham University Database.

Addessi, E., A. Mancini, L. Crescimene, C. Pasoa-Schioppa, and E. Visalberghi. "Preference Transivity and Symbolic Representation in Capuchin Monkeys (*Cebus apella*)." *PLOS ONE*, 3, no. 6 (2008): 1–8.

Addessi, E., and L. Visalberghi. "Do Capuchin Monkeys (*Cebus apella*) Use Tokens as Symbols?" *Proceedings of the Royal Society of London*, August 14, 2007. https ://doi.org/10.1098/rspb.2007.0726.

Alamgià, R. *Monumenta cartographia vaticana issu Pii XII P.M. consilio et opera procuratorum Bibliothecae Apostalicae Vaticana.* Vol. 1. Vatican City: Biblioteca Apostolica Vaticana, 1944.

"Apollo 14 Hike to Cone Crater." NASA Visualization Studio, February 8, 2021. https://svs.gsfc.nasa.gov/4883.

Arbel, B. "Maps of the World for Ottoman Princes? Further Evidence and Questions Concerning 'The *Mappamondo* of Hajji Ahmed.'" *Imago Mundi* 54, no. 1 (2002): 19–29.

Aubert, M., P. Setiawan, A. Brumm, A. Oktaviana, A. Oktaviana, P. H. Sulistyarto, E. W. Saptomo, B. Istiawan, and T. A. Ma'rifat. "Paleolithic Cave Art in Borneo." *Nature*, 564 (2018): 254–257.

Aujac, G. "The Foundations of Theoretical Geography in Archaic and Classical Greece." In *The History of Cartography: Cartography in Prehistoric, Ancient, and Medieval Europe and the Mediterranean*, vol. 1, edited by J. B. Harley and D. Woodward, 130–47. Chicago: University of Chicago Press, 1987.

———. "Greek Cartography in the Early Roman World." In *The History of Cartography: Cartography in Prehistoric, Ancient, and Medieval Europe and the Mediterranean*, vol. 1, edited by J. B. Harley and D. Woodward, 161–76. Chicago: University of Chicago Press, 1987.

———. "The Growth of Empirical Cartography in Hellenistic Greece." In *The History of Cartography: Cartography in Prehistoric, Ancient, and Medieval Europe and the Mediterranean*, vol. 1, edited by J. B. Harley and D. Woodward, 148–60. Chicago: University of Chicago Press, 1987.

Belisle, P. D. "Overview of Camaldolese History and Spirituality." In *The Privilege of Love: Camaldolese Benedictine Spirituality*, edited by P. D. Belisle, 3–26. Collegeville, MN: Liturgical Press, 2002.

Blaut, J.M., D. Stea, C. Spenser and M. Blades. "Mapping as a Cultural and Cognitive Universal." *Annals of the Association of American Geographers*, 93 (2003): 163–185.

Blackmore, M. "From Way-Finding to Map-Making; The Spatial Information
    Fields of Aboriginal Peoples." *Progress in Human Geography*, 5, vol. 1 (1981):
    1–24.

Bouzouggara, A., N. Barton, M. Vanhaeren, F. d'Errico, S. Collcutt, T. Highham,
    and R. Hodge. 82,000-year-old Shell Beads from North Africa and Implications
    for the Origins of Modern Human Behavior." *Proceedings of the National Academy
    of Sciences*, 104 (2007): 1964–1969.

Bowd, S. D. *Reform before the Reformation: Vincenzo Querini and the Religious
    Renaissance in Italy*. Leiden: Brill, 2002.

Boysen, S. "Counting in chimpanzees: Nonhuman principals and emergent
    properties of number." In *The Development of Numerical Competence: Animal
    and Human Model*, edited by Boysen, S. and E. J. Capadi, 39–60. New York:
    Psychology Press.

Brentjes, S. "Fourteenth-Century Portolan Charts: Challenges to Our
    Understanding of Cross-Cultural Relationships in the Mediterranean and
    Black Sea Regions and of (Knowledge?) Practices of Chart-Makers." *Journal of
    Transcultural Medieval Studies*, 2, no. 1 (2015): 79–122.

Brooks, A., J. E. Yellen, R. Potts, A. K. Behrensmeyer, A. L. Deino, D. E. Leslie,
    and S. H. Ambrose. "Long-distance stone transport and pigment use in the
    earliest Middle Stone Age." *Science* 360, vol. 6384 (2018): 90–94.

Brotton, J. *A History of the World in 12 Maps*. New York: Penguin (2012).

———. *Great Maps*. Washington, DC: DK Smithsonian, 2014.

Call, J. and M. Tomasello. "Reasoning and thinking in nonhuman primates." In *The
    Cambridge Handbook of Thinking and Reasoning* edited by Holyoak, K. and R.
    Morrison, 607–625. Cambridge: Cambridge University Press, 2006.

"Camaldolese Nuns of Poppi." https://www.camaldolesidipoppi.it/en/.

"Camaldoli, A Hermitage and a Monastery." https://www.ilbelcasentino.it
    /camaldoli-en.html.

Campagnol, I. *Forbidden Fashions: Invisible Luxuries in Early Venetian Convents*.
    Lubbock: Texas Tech University Press, 2014.

Campbell, T. "Portolan Charts from the Late Thirteenth Century to 1500." In *The
    History of Cartography: Cartography in Prehistoric, Ancient, and Medieval Europe
    and the Mediterranean*, vol. 1, edited by J. B. Harley and D. Woodward, 371–
    463. Chicago, University of Chicago Press, 1987.

Cattaneo, A. "Building a World Unified by Maritime Networks: Fra Mauro's Mappa
    Mundi between Venice and Lisbon, ca. 1450." YouTube, December 18, 2015.
    https://www.youtube.com/watch?v=0n8XGZ_mfMk.

———. *Fra Mauro's Mappa Mundi and Fifteenth-Century Venice*. Turnhout, Belgium:
    Brepols, 2011.

———. "God in His World: The Earthy Paradise in Fra Mauro's 'Mappa Mundi'
    Illuminated by Leonardo Bellini." *Imago Mundi* 55 (2003): 97–102.

———. "La mappa mundi di Fra Mauro, l'idea di oceano e le direzioni di navigazione
    all'alba dell'espansioner. In *Mundus novus: Amerigo Vespucci e i metodi della ricerca
    storico-geografica*, edited by A. D'Ascenzo, 109–21. Genoa, Italy: Brigati, 2004.

Cazzagon, G. *La Chiesa de San Michele in Isola*. Venice: Grafiche Veneziane, 2021.

Cheon, S. A., A. Nanetti, and M. Filippov. "Digital Maps and Automatic Narratives for the Interactive Global Histories." *Asian Review of World Histories* 4, no. 1 (2016): 83–123.

"The Colorful Medieval World: Mixing Ancient Colors." Insanitek, May 10, 2019. https://insanitek.net/the-colorful-medieval-world-mixing-ancient-ink-colors/.

Cotton, J. "The Churches of Venice." n.d. http://churchesofvenice.com/ven churches.htm.

Cross, F. L., and E. A. Livingston, eds. "Camaldolese." In *The Oxford Dictionary of the Christian Church*. 3rd ed. Oxford: Oxford University Press, 2005.

Dalché, P. G. "The Reception of Ptolemy's Geography (End of the Fourteenth to Beginning of the Sixteenth Century)." In *The History of Cartography*, vol. 3, part 1, edited by D. Woodward, 285–364. Chicago: University of Chicago Press, 2007.

Davenport, W. "Marshall Islands Navigational Charts." *Imago Mundi*, 15, vol. 1 (1960): 19–26.

Deacon, T. *The Symbolic Species*. New York: W. W. Norton & Sons, 1998.

"The Decolonial Atlas." https://decolonialatlas.wordpress.com/.

DeHutorowicz, H. "Maps of Primitive Peoples." *Bulletin of the American Geographical Society of New York*, 43, vol. 1 (1911): 669–680.

DeLoach, J. "Becoming symbol-minded." *Science* 8, no. 2 (2004): 66–70.

d'Errico, F., C. Hershilwood, M. Vanhaeren, and K. van Niekerk. "*Nassarius kraussianus* Shell Beads from Blombos Cave: Evidence for Symbolic Behavior in the Middle Stone Age." *Journal of Human Evolution*, 48, vol. 1 (2005): 3–24.

di Robilant, A. *Irresistible North: From Venice to Greenland on the Trail of the Zen Brothers*. New York: Alfred A. Knopf, 2011.

Dilke, O. A. W. "The Culmination of Greek Cartography in Ptolemy." In *The History of Cartography: Cartography in Prehistoric, Ancient, and Medieval Europe and the Mediterranean*, vol. 1, edited by J. B. Harley and D. Woodward, 177–200. Chicago: University of Chicago Press, 1987.

Edson, E. "Maps in Context: Isidore, Orosius, and the Medieval Image of the World." In *Cartography in Antiquity and the Middle Ages*, edited by J. A. Talbert and R. W. Unger, 219–36. Leiden: Brill, 2008.

———. *The World Map, 1300–1492: The Persistence of Tradition and Transformation*. Baltimore: Johns Hopkins University Press, 2007.

Emiralioglu, P. "Cartography and the Ottoman Imperial Project in the Sixteenth Century." In *Imperial Geographies in Byzantine and Ottoman Space*, edited by S. Angelov, S. Bazzaz, and Y. Batski, 114–148. Cambridge, MA: Harvard University Press, 2013.

"Engineering Historical Memory." https://engineeringhistoricalmemory.com/.

Falchetta, P. *Fra Mauro's World Map: A History*. London: Imago, 2013.

———. *Fra Mauro's World Map with a Commentary and Translation of the Inscriptions*. Turnhout, Belgium: Brepols, 2006.

———. "Manuscript No. 10057 in the Biblioteca Marciana, Venice: A Possible Source for the Catalan Atlas?" *Imago Mundi* 46 (1994): 19–28.

———. *Marinai, mercanti, cartografi, pittori: ricerche sulla cartografia nautica a Venezia, sec. XIV-XV.* Venice; Ateneo Veneto, 1995.

———. *Storia del Mappamondo di Fra' Mauro: Con la Trascrizione Integrale del Testo.* London: Imago, 2013.

Ferrar, M. J. "The Ptolemaic Influence upon Andreas Bianco; Atlante Nautico, 1436." Cartography Unchained, October 2022. https://www.cartography unchained.com/chab2/.

"Fra Mauro's Map of the World (Dated 26 August 1460)." Engineering Historical Memory, n.d. https://engineeringhistoricalmemory.com/.

"Fra Mauro's Mappamundi." NASA, Landsat Science, January 17, 2014. https ://landsat.gsfc.nasa.gov/article/fra-mauros-mappamundi/.

"Fra Mauro's World Map." Museo Galileo, n.d. https://mostre.museogalileo.it /framauro/en/historical-context-m/the-world-map/from-the-mediterranean-to -the-indian-ocean.html.

Gallez, P. "Walsperger and His Knowledge of the Patagonian Giants." *Imago Mundi*, 33 (1981): 91–93.

Garfield, S. *On the Map: A Mind-Expanding Exploration of the Way the World Looks.* New York: Avery, 2013.

Garner, R. "Landsat Overview." NASA, December 9, 2021. https://www.nasa.gov /mission_pages/landsat/overview/index.html.

Gow, A. "Gog and Magog on Mappaemundi and Early Printed World Maps: Orientalizing Ethnography in the Apocalyptic Tradition." *Journal of Early Modern History* 2, no. 1 (1998): 61–88.

Hale, R. "Koinonia: The Privilege of Love." In *The Privilege of Love: Camaldolese Benedictine Spirituality*, edited by P. D. Belise, 99–114. Collegeville, MN Liturgical Press, 2002.

"Hampi: City of Ruins." Incredible !ndia, n.d. https://www.incredibleindia.org /content/incredibleindia/en/destinations/hampi.html.

Harley, J. B. "The Map and the History of the Development of Cartography." In *The History of Cartography: Cartography in Prehistoric, Ancient, and Medieval Europe and the Mediterranean*, vol. 1, edited by J. B. Harley and D. Woodward, 1–42. Chicago: University of Chicago Press, 1987.

Harley, J. B., and D. Woodward. *The History of Cartography: Cartography in Prehistorical, Ancient, and Medieval Europe and the Mediterranean.* Vol. 1. Chicago: University of Chicago Press, 1987.

Harvey, P. D. "Medieval Maps: An Introduction." In *The History of Cartography: Cartography in Prehistoric, Ancient, and Medieval Europe and the Mediterranean*, vol. 1, edited by J. B. Harley and D. Woodward, 283–85. Chicago: University of Chicago Press, 1987.

———. *Mappa Mundi: The Hereford Map.* Toronto: University of Toronto Press, 1996.

———. "The Sawley Map and Other World Maps in Twelfth-Century England." *Imago Mundi* 49, no. 1 (1997): 33–42.

Henshilwood, C., F. d'Errico, K. van Niekerk, L Dayet, A. Queffeclec, and L.
   Pollarolo. "An Abstract Drawing from the 73,000-year-old Level of Blombos
   Cave." *Nature*, 562 (2003): 115–118.
Henshilwood, C., F. d'Errico, M. Vanhaeren, K. van Niekerk, and Z. Jacobs.
   "Middle Stone Age Shell Beads from South Africa." *Science*, 304, vol. 5669
   (2004): 404–405.
Hildago, R. J., R. Córdoba, P. Nabais, V. Silva, M. J. Melo, F. Pina, N. Teixeira, and
   V. Freita. "New Insights into Iron-Gall Inks through the Use of Historically
   Accurate Reconstructions." *Heritage Science* 6, no. 63 (2018): 1–15.
Hintz, Charlie. "Death and Burial in Venice: What Does the Floating City Do With
   Its Dead?" Cult of Weird, n.d. https://www.cultofweird.com/about/.
"History of the Venetian Civic Cemetery." Venetian Protestant Cemetery, 2016.
   http://www.veniceprotestantcemetery.com/en/history.
Hoffmann, D. L., D. E. Angelucci, A. Villaverde, J. Zapata, and J. Zilhão.
   "Symbolic Use of Marine Shells and Mineral Pigments by Iberian Neanderthls."
   Science Advances, 4, vol. 2 (2018): 5255.
Hoffmann, D. L., C. D. Standish, M. Garcia-Diez, P. B. Pettit, J. A. Milton,
   J. J. Zilhão, and J. J. Alcolea-Gonzáles. "U-Th Dating of Carbonate Crusts
   Reveals Neanderthal Origin of Iberian Cave Art." *Science*, 359, vol. 6378 (2018):
   912–915.
"How to Make Ink in the Middle Ages." Mediavalists.net, n.d. https://www
   .medievalists.net/2015/09/how-to-make-ink-in-the-middle-ages/.
Howard, D. *The Architectural History of Venice*. 2nd ed. New Haven, CT: Yale
   University Press, 2002.
Hublin, J., A. Ben-Ncer, S. E. Baily, S. E. Freidline, S. Neubauer, M. M. Skinner,
   and I. Bergman. "New fossils from Jebel Irhoud, Morocco and the pan-
   American origin of *Homo sapiens*." *Nature*, 546 (2017): 289–292.
"Inuit Cartography." The Decolonial Atlas, https://decolonialatlas.wordpress
   .com/2016/04/12/inuit-cartography/.
"Islamic Cosmology and Astronomy: Ibrahim Hakki's Maarifetname." World
   Bulletin, September 20, 2010. https://worldbulletin.dunyabulteni.net/art
   -culture/wellknown-ottoman-book-marifetname-translated-into-english
   -h64138.html.
Izard, V., C. Sann, S. Spelke, ad A. Streri. "Newborn infants perceive abstract
   numbers." *Proceedings of the National Academy of Sciences* 106, no. 25, (2009):
   10382-10385.
Jabr, F. "Hunting for the origins of symbolic thought." *New York Times*, December 12,
   2014. https://www.nytimes.com/2014/12/07/magazine/hunting-for-the-origins
   -of-symbolic-thought.html?searchResultPosition=1
Janmaat, K., C. Boesch, R. Byrne, C. Chapman, Z. B. Goné Bi, J. S. Head,
   M. M. Robbins, R. Wrangham, and L. Polansky. "Spatio-temporal complexity
   of chimpanzee food: How cognitive adaptations can counteract the ephemeral
   nature of ripe fruit." *American Journal of Primatology*, 78, vol. 6 (2016): 626–645.

Johns, J., and E. Savage-Smith. "The Book of Curiosities: A Newly Discovered Series of Islamic Maps." *Imago Mundi* 55, no. 5 (2003): 7–24.

Kagan, R. L., and B. Schmidt. "Maps and the Early Modern State: Official Cartography." In *The History of Cartography*, vol. 3, part 1, edited by D. Woodward, 661–79. Chicago: University of Chicago Press, 2007.

Karamustafa, A. T. "Introduction to Islamic Maps." In *The History of Cartography*, vol. 2, book 1, edited by J. B. Harley & D. Woodward, 3–11. Chicago: University of Chicago Press, 1992.

Kitchen, R. M. "Cognitive maps: What they are and why study them?" *Journal of Environmental Science*, 14 (1994): 1–19.

Kurlansky, M. *Cod: The Fish That Changed the World*. New York: Penguin, 1997.

———. *Paper: Paging through History*. New York: W. W. Norton, 2016.

———. *Salt: A World History*. London: Walker Publishing Company, 2002.

Lane, F. C. "Venetian Shipping during the Commercial Revolution." *American Historical Review* 38, no. 2 (1933): 219–39.

———. *Venetian Ships and Shipbuilders of the Renaissance*. Baltimore: Johns Hopkins Press, 1934.

———. *Venice: A Maritime Republic*. Baltimore: Johns Hopkins University Press, 1973.

Larner, J. *Marco Polo and the Discovery of the World*. New Haven, CT: Yale University Press, 1999. Cambridge: Cambridge University Press, 1976.

Laven, M. *The Virgins of Venice: Broken Vows and Cloistered Lives in the Renaissance Convent*. New York: Viking, 2003.

Leach, E. *Culture and Communication: The Logic by Which Symbols are Connected*. Cambridge: Cambridge University Press, 1976.

"Least Distorted Equal Area World Map 2016." Good Design® Awards, n.d. https://www.good-designawards.com/award-details.html?award=31558.

Ledyard, G. "Cartography in Korea." In *The History of Cartography: Cartography in the Traditional East and Southeast Asian Societies*, vol. 2, book 2, edited by J. B. Harley and D. Woodward, 235–344. Chicago: University of Chicago Press, 1994.

Lee, R. B., and R. Daly. *The Cambridge Encyclopedia of Hunters and Gatherers*. Cambridge: Cambridge University Press, 1999.

Leporace, G. T. *Il Mappamundo Di Fra Mauro*. Rome: Istituto poligrafico dello State, Libreria dello Stato, 1956.

Lewis, M. "The origins of cartography." In *The History of Cartography*, vol. 1, edited by J. Harley and D. Woodward, 50–53. Chicago, University of Chicago Press, 1987.

Life-Size 1871 Photograph of the Fra Mauro Map of the World. Venice: Carlo Naya, ca. 1871. https://www.abebooks.com/Life-size-1871-photograph-Mauro-map-world/30750493331/bd.

Lyons, H. "The Sailing Charts of the Marshall Islanders." *Geographical Journal*, 72, vol. 4 (1928): 325–327.

Macron, S. "Il mappamundo di Fra Mauro e Leonardo Bellini." In *per l'arte da Venezia all'Europa. Studi in onore di Giuseppe Maria Pilo*, edited by M. Piantoni and L. deRossi, 103–8). Gorizia, Italy: Edizioni della Laguna, 2001.

———. "Leonardo Bellini and Fra Mauro's World Map: The Earthly Paradise." In *Fra Mauro's World Map*, edited by P. Falchetta, 135–61. Turnhout, Belgium: Brepols, 2006.

Magno, A. M. *Dei libri; Quando Venezia ha fatto leggere il mondo [Bound in Venice: The Serene Republic and the Dawn of the Book]*. New York: Europa Editions, 2013.

Mai, T. "Global Positioning System History." NASA, August 7, 2017. https://www .nasa.gov/directorates/heo/scan/communications/policy/GPS_History.html.

Mappamundo Drawn by Fra Mauro, 1459 [Cartographic Material]. National Library of Wales, ca. 1869. https://discover.library.wales/primo-explore/fulldisplay?docid =44NLW_ALMA21731596320002419&context=L&vid=44WHELF_NLW _NUI&lang=en_US&search_scope=LSCOP_INLIBRARY&adaptor =Local%20Search%20Engine&tab=tab4&query=any,contains,Fra%20Mauro.

Maqbul Ahmad, S. "Cartography of al-Sharīf al-Idrīsī." In *The History of Cartography: Cartography in the Traditional Islamic and South Asian Societies*, vol. 1, book 2, edited by J. B. Harley and D. Woodward, 136–72. Chicago: University of Chicago Press, 1992.

Matsuzawa, T. *Primate Origins of Human Cognition and Behavior*. New York: Springer, 2001.

———. "Symbolic Representation of Numbers in Chimpanzees." *Current Opinion in Neurobiology* 19 (2009): 92–98.

Mauntel, C. "Fra Mauro's View of the Boring Question of Continents." *Peregrinations: Journal of Medieval Art and Architecture* 6, no. 3 (2018): 54–77.

McGreggor, J. H. S. *Venice from the Ground Up*. Cambridge, MA: Belknap Press, Harvard University Press, 2006.

McPhail, C. K. "Reconstructing Eratosthenes' Map of the World: A Study in Source Analysis." Master's thesis, University of Otago, 2011. https://ourarchive.otago .ac.nz/bitstream/handle/10523/1713/McPhailCameron2011MA.pdf.

"Medieval Age (500–1400)." Pigments through the Ages, n.d. http://www.web exhibits.org/pigments/intro/medieval.html.

Mercator, G. "World Map on Double Cordiform Projection." Library of Congress, 1538. https://www.loc.gov/item/2021668435/.

Millard, A. R. "Cartography in the Ancient Near East." In *The History of Cartography: Cartography in Prehistoric, Ancient, and Medieval Europe and the Mediterranean*, vol. 1, edited by J. B. Harley and D. Woodward, 107–16. Chicago: University of Chicago Press, 1987.

"Monastery Stays." https://www.monasterystays.com/.

Morris, J. *The World of Venice*. New York: Pantheon, 1960.

Morse, V. "The Role of Maps in Later Medieval Society: Twelfth to Fourteenth Century." In *The History of Cartography*, vol. 3, part 1, edited by D. Woodward, 25–52. Chicago: University of Chicago Press, 2007.

Nanetti, A. "Waterways Connecting Peoples of the World: A Presentation of the EHM Application for Fra Mauro's Mappa Mundi as a Virtual Laboratory for Investigating the Maritime Silk Road Discourse in the Digital Time Machine." In *Venezia e Il Senso Del Mare: Percezioni e Rappresentazioni*, edited by

M. Aymard and E. Orlando, 161–250. Venice: Istituto Veneto di Scienze, Lettere ed Arti, 2022a.

———. *Computational Engineering of Historical Memory: With a Showcase on Afro-Eurasia (ca 1100–1500 CE)*. Abingdon, UK: Routledge, 2022b.

Nanetti, A., A. Cattaneo, S. A. Cheong, and C.-Y. Lin. "Maps as Knowledge Aggregators: From Renaissance Italy Fra Mauro to Web Search Engines." *Cartographic Journal* 52, no. 2 (2015): 159–67.

Nanetti, A., and D. Benvenuti. "Animation of Two-Dimensional Pictorial Works into Multipurpose Three-Dimensional Objects: The Atlas of the Ships of the Known World Depicted in the 1460 Fra Mauro's Mappa Mundi as a Showcase." *Scientific Research and Information Technology* 9, no. 2 (2019): 29–46.

Nanetti, A., and D. Benvenuti. "Engineering Historical Memory and the Interactive Exploration of Archival Documents: The Online Application for Pope Gregory X's Privilege for the Monastic Community of Mount Sinai (1274) as a Prototype." *Umanistica Digitale*, September 9, 2021. https://umanisticadigitale .unibo.it/article/view/12567.

Nanetti, A., and S. A. Cheong. "The World as Seen from Venice (1205–1533) as a Case Study of Scalable Web-Based Automatic Narratives for Interactive Global Histories." *Asian Review of World Histories* 4, no. 1 (2016): 3–34.

National Library of Australia. *Mapping Our World: Terra Incognita to Australia*. YouTube, November 7, 2013. https://www.youtube.com/watch?v=WjA3mfq M09s&ab_channel=VisitCanberra.

National Library of Australia. "From Venice to the National Library of Australia: The Fra Mauro Story." YouTube, March 13, 2014. https://www.youtube.com /watch?v=giDBZd8-k0c&ab_channel=Antoniocanovazardo.

"New Camadoli Hermitage." https://www.contemplation.com/.

Norris, R. and B. Harney "Songlines and Navigation in Wardaman and Other Australian Aboriginal Cultures." *Journal of Astronomical History and Heritage*, 17, vol. 2 (2014): 1–15.

O'Doherty, M. "Fra Mauro's World Map (c. 1448–1459): Mapping, Meditation, and the Indian Ocean World in the Early Renaissance." *Wasafiri* 26, no. 2 (2011): 30–36.

Premack, D. and A. Premack *The Mind of an Ape*. New York: Norton, 1983.

Pérez, S. *The Beatus Maps: The Revelation of the World in the Middle Ages*. Burgos, Spain: Siloé, 2014.

"The Peters Projection Map." Oxford Cartographers, n.d. https://www.oxford cartographers.com/our-maps/peters-projection-map/.

Piaget, J. *Origins of Intelligence in the Child*. Abingdon, UK: Routledge, 1936.

Piaget, J. and B. Inhelder. *The Child's Conception of Space*. London: London, 1948.

Pinto, K. "Medieval Islamic Maps: An Exploration." Chicago: University of Chicago Press, 2016.

Pizzati, L. *Venetian-English, English-Venetian: When in Venice Do as the Venetians*. Bloomington, IN: AuthorHouse, 2007.

Polo, M. *The Travels of Marco Polo*. Barnes & Noble Publishing, 2012.

Povoledo, E. "A Chance to Spend 99-Plus Years in Venice (in the Afterlife)." *New York Times*, May 22, 2018. https://www.nytimes.com/2018/05/21/world/europe /venice-cemetery-auction.html.

Ramusio, G. *Delle navigationi e viaggi*. Venice: Appreffo i Giunti, 1606.

Rappaport, Y., and E. Savage-Smith. *An Eleventh-Century Guide to the Universe*. Leiden, Brill, 2014.

Rappaport, Y., and E. Savage-Smith. "Medieval Islamic View of the Cosmos: The Newly Discovered Book of Curiosities." *Cartographic Journal* 41, no. 3 (2004): 253–59.

Ratti, A. "A Lost Map of Fra Mauro Found in a Sixteenth Century Copy." *Imago Mundi* 40 (1988): 77–85.

Rodriguex-Hodalgo, A., J. I., Morales, A. Cebrià, A. Courtenay, G. Fernández-Marchena, G. Garcia-Argudo, and J. Marin. "The Chatelperronian Neanderthals of Cova Foradada (Calafell, Spain) used imperial eagle phalanges for symbolic purposes." *Science Advances*, 5, vol. 11 (2018): 1984.

Routley, N. "Mapped: Where Are the World's Ongoing Conflicts Today." Visual Capitalist, October 4, 2021. https://www.visualcapitalist.com/mapped -where-are-the-worlds-ongoing-conflicts-today/.

Rundstrum, R. "A Cultural Interpretation of Inuit Map Accuracy." *Geographic Review*, 80, vol. 2 (1990): 155–168.

Ryder, M. L. "Parchment—Its History, Manufacture and Composition." *Journal of the Society of Archivists* 2, no. 9 (1964): 391–99.

Sample, I. "Earliest Known Drawing Found on Rock Art in South African Cave." *The Guardian*, September 12, 2018. https://www.theguardian.com/science/2018 /sep/12/earliest-known-drawing-found-on-rock-in-south-african-cave.

Savage-Rumbaugh, S., S. G. Shaker and T. T. Taylor. *Apes, Language, and the Human Mind*. Oxford: Oxford University Press, 1998.

Savoy, D. "A Ladder of Calmadolite Salvation: The Facade of San Michele in Isola." *Athanor* 20 (2002): 34–41.

Scafi, A. "Mapping Eden: Cartographies of Earthly Paradise." In *Mappings*, edited by D. Cosgrove, 51–70. London: Reaktion Books, 1999.

Scardia, G., W. A. Neves, I. Tattersall, and L. Blumrich. (2021). "What Kind of Hominin First Left Africa?" *Evolutionary Anthropology* 30, no. 2 (2021): 122–27.

Scharping, N. "Finally, a World Map That Doesn't Lie." *Discover*, November 3, 2016. https://www.discovermagazine.com/environment/finally-a-world -map-that-doesnt-lie.

"Sea Rise Levels and the Fate of Coastal Cities." Google Earth, n.d. https://earth .google.com/web/@-77.06887022,164.39811389,-10.6422799a,787298 .47509228d,35y,166.67160529h,72.48516064t,0r/data=CjESLxIgNzJlM 2QwZWU3NGMyMTFlODhjMWNiZjg2OTQ1ZTVlZWMiC 3ZveV9wb2ludF8x.

Sheehan, K. E. "Aesthetic Cartography: The Cultural Function of Portolan Charts from 1300 to 1700." *Imago Mundi* 65, no. 1 (2013): 133–35.

Silverberg, R. *The Realm of Prester John*. Athens: Ohio University Press, 1972.

Simon, Z. A. "Secrets of South Asia from Fra Mauro's (1459) to Later Maps."
    Cartographica 52, no. 3 (2017): 263–87.

Small, M. F. *Inventing the World: Venice and the Transformation of Western Civilization.*
    New York: Pegasus Books, 2020.

Smith, C. D. "The Emergence of 'Maps' in European Rock Art: A Prehistoric
    Preoccupation with Place." *Imago Mundi* 34, no. 1 (1982): 9–25.

———. "Prehistoric Maps and the History of Cartography." In *The History of
    Cartography*, vol. 1, edited by J. B. Harley and D. Woodward, 45–49. Chicago:
    University of Chicago Press, 1987.

Smith, D. *The State of the World Atlas.* New York: Penguin, 2021.

Snyder, J. P. "Map Projections in the Renaissance." In *The History of Cartography*,
    vol. 3, part 1, edited by D. Woodward, 365–81. Chicago: University of Chicago
    Press, 2007.

Sokol, J. "Can This New Map Fix Our Distorted Views of the World?" *New York
    Times*, February 24, 2021. https://www.nytimes.com/2021/02/24/science/new
    -world-map.html.

Special Collection Conservation Unit of the Preservation Department of Yale
    University Library. *Medieval Manuscripts: Some Ink and Pigment Recipes.* Booklet,
    2012. https://travelingscriptorium.files.wordpress.com/2012/03/scopa-recipes
    -booklet_web.pdf.

Spennemann, D. "Traditional and Nineteenth Century Communication Patterns in
    the Marshall Islands." *Micronesian Journal of Humanities and Social Sciences*, 4,
    vol. 1 (2005): 25–52.

St. Fleur, N. "Oldest known drawings by human hands discovered in South African
    cave." *New York Times*, September 12, 2018. https://www.nytimes.com/2018/09
    /12/science/oldest-drawing-e.

Steins, J. J. "Woodblock Printing." January 19, 2010. https://www.johnsteins.com
    /woodblock-printing.html/.

Stewart, J. "The History of the Color Red: From Ancient Painting to Louboutin
    Shoes." My Modern Met, September 26, 2018. https://mymodernmet.com
    /shades-of-red-color-history/.

Tattersall, I. "An Evolutionary Framework of the Acquisition of Symbolic Cognition
    by *Homo sapiens.*" *Comparative Cognition and Behavior Reviews*, 3 (2008): 99–114.

Thrower, N. J. W. *Maps and Civilization: Cartography in Culture and Society.* Chicago:
    University of Chicago Press, 1972.

———. *Maps and Civilization: Cartography in Culture and Society.* 3rd ed. Chicago:
    University of Chicago Press, 2008.

Tibbets, G. R. "The Balkhī School of Geographers." In *The History of Cartography:
    Cartography in the Traditional Islamic and South Asian Societies*, vol. 2, book 1,
    edited by J. B. Harley and D. Woodward, 108–36. Chicago: University of
    Chicago Press, 1992.

———. "The Beginnings of a Cartographic Tradition." In *The History of Cartography:
    Cartography in the Traditional Islamic and South Asian Societies*, vol. 2, book 1,

edited by J. B. Harley and D. Woodward, 90–107. Chicago: University of Chicago Press, 1992.

Tolman, E. "Cognitive maps in rats and men." *The Psychological Review*, 55, vol. 44 (1948): 189–208.

"Tools of the Trade Part 1, Brushes." *Medieval Colours* (blog), March 24, 2010. https://medievalcolours.blogspot.com/2010/03/tools-of-trade-part-1-brushes -and-pens.html.

Turconi, C. "La mappa di Bedolina nel quadro'arte ruestre della Valmonica." *Notizie Archeologiche* (1985): 85–113.

Turconi, C. "La mappa de Bedolina: Il suo signato nel quadro dell'arte rupestre camuna dell'eta' del Ferro." In *Archaeologia e arte repestre L'Europa, Le Alpi La Valcamonica*, 239–243. Milano, 2001.

Turley, T. "Romuald of Ravenna, Saint (c. 952–1027)." In *Key Figures in Medieval Europe: An Encyclopedia*, edited by R. K. Emmerson and N. S. Clayton-Emmerson, 578–79. Abingdon, UK: Routledge, 2006.

Tylén, K., R. Fusaroli, S. Rojo, K. Heimann, N. Fay, N. N. Johannes, F. Riede, and M. Lombard. "The Evolution of Early Symbolic Behavior in *Homo sapiens*." *Proceedings of the National Academy of Science*, 117, vol. 9 (2020): 4578–4584.

Uri, J. "50 Years Ago: Apollo 14 Lands at Fra Mauro." NASA, February 4, 2021. https://www.nasa.gov/feature/50-years-ago-apollo-14-lands-at-fra-mauro.

van Donzel, E. S., and A. Schmidt, eds. *Gog and Magog in Early Syriac and Islamic Sources: Salaam's Quest of Alexander's Wall*. Leiden, Brill, 2009.

van Duzer, C. *Sea Monsters and Medieval and Renaissance Maps*. London: British Library, 2013.

———. *Sea Monsters on Medieval and Renaissance Maps*. London: British Library, 2014.

ven den Hoonaard, W. C. *Map Worlds: History of Women in Cartography*. Waterloo, Ontario: Wilfrid Laurier University Press 2013.

"Venice's 'Bricole': Sentinels of the Lagoon." Pieces of Venice, January 24, 2020. https://piecesofvenice.com/en/venice-s-bricole-sentinels-of-the-lagoon/.

Vigilucci, L. *Camaldoli: A Journey into Its History and Spirituality*. Naperville, IL: Source Books, 1988.

Vogel, K. A. "Fra Mauro and the Modern Globe." *Globe Studies* 57/58 (2011): 81–92.

Wetrem, S. D. *The Hereford Map: A Transcription and Translation of the Legends with Commentary*. Turnhout, Belgium: Brepols, 2001.

Whitmore, J. K. "Cartography in Vietnam." In *The History of Cartography: Cartography in the Traditional East and Southeast Asian Societies*, vol. 2, book 2, edited by J. B. Harley and D. Woodward, 478–508. Chicago: University of Chicago Press, 1994.

Wilford, J. N. *The Mapmakers*. New York: Random House, 1981.

Williams, D. R. "The 'Moon Trees.'" NASA, November 1, 2022. https://nssdc.gsfc .nasa.gov/planetary/lunar/moon_tree.html.

Wills, G. *Venice, Lion City: The Religion of Empire*. New York: Simon & Schuster, 2001.

Wise, D. "Primitive cartography in the Marshall Islands." *Cartographica,* 13, vol. 1 (1976): 11–20.

Winter, H. "The Fra Mauro Portolan Chart in the Vatican." *Imago Mundi* 16 (1962): 17–28.

Wong, K. "Did Neanderthals Think Like "Us"?" *Scientific American*, 203, vol. 6 (2010): 72–75.

Wood, D. "The Fine Line Between Mapping and Map Making." *Cartographica*, 40, vol. 4 (1993): 50–60.

Woodward, D. "Medieval Mappa Mundi." In *The History of Cartography*, vol. 1, edited by J. B. Harley and D. Woodward, 203–55. Chicago: University of Chicago Press, 1987.

———. "Reality, Symbolism, Time, and Space in Medieval World Maps." *Annals of the Association of American Geographers* 75, no. 4 (1985): 510–21.

"Worldmapper." https://worldmapper.org/maps/.

Wraggs Sykes, R. *Kindred: Neanderthal Life, Love, Death, and Art*. New York: Bloomsbury Sygma, 2020.

Wright, J. K. *The Leardo Map of the World, 1452 or 145*. Vol. 4. Concord, NH: Rumford Press, 1928.

Yuhas, A. (2021, September 30). "Yale Says Its Vinland Map, Once Called a Medieval Treasure, Is Fake." *New York Times*, September 30, 2021. https://www.nytimes.com/2021/09/30/us/yale-vinland-map-fake.html.

Zilhão, J. "Symbolic Use of Marine Shell and Mineral Pigments by Iberian Neanderthals." *Proceedings of the National Academy of Sciences*, 107. vol. 3 (2010): 1023-1028.

Zorzi, A. *Venice 697–1797: A City, A Republic, An Empire*. Milan: Arnoldo Mondadori Editore, 1999.

Zurla, P. *Il Mappamondo di Fra Mauro*. Rimini: Imago, 1806.

Index